# Discovering Business Intelligence Using MicroStrategy 9

Leverage the power of the MicroStrategy platform to design, develop, secure, and share Business Intelligence reports and dashboards

**Nelson Enriquez**

**Samundar Singh Rathore**

BIRMINGHAM - MUMBAI

# Discovering Business Intelligence
# Using MicroStrategy 9

Copyright © 2013 Packt Publishing

All rights reserved. No part of this book may be reproduced, stored in a retrieval system, or transmitted in any form or by any means, without the prior written permission of the publisher, except in the case of brief quotations embedded in critical articles or reviews.

Every effort has been made in the preparation of this book to ensure the accuracy of the information presented. However, the information contained in this book is sold without warranty, either express or implied. Neither the authors, nor Packt Publishing, and its dealers and distributors will be held liable for any damages caused or alleged to be caused directly or indirectly by this book.

Packt Publishing has endeavored to provide trademark information about all of the companies and products mentioned in this book by the appropriate use of capitals. However, Packt Publishing cannot guarantee the accuracy of this information.

First published: December 2013

Production Reference: 1171213

Published by Packt Publishing Ltd.
Livery Place
35 Livery Street
Birmingham B3 2PB, UK.

ISBN 978-1-78217-004-4

www.packtpub.com

# Credits

**Authors**

Nelson Enriquez

Samundar Singh Rathore

**Reviewers**

Chael Christopher

Steven Zagoudis

**Acquisition Editor**

Kevin Colaco

**Commissioning Editors**

Neil Alexander

Neha Nagwekar

**Technical Editors**

Shashank Desai

Manal Pednekar

**Copy Editors**

Roshni Banerjee

Sarang Chari

Dipti Kapadia

Karuna Narayanan

Shambhavi Pai

**Project Coordinator**

Michelle Quadros

**Proofreader**

Cathy Cumberlidge

**Indexer**

Mariammal Chettiyar

**Production Coordinator**

Pooja Chiplunkar

**Cover Work**

Pooja Chiplunkar

**Cover Image**

Valentina D'silva

# About the Authors

**Nelson Enriquez** is a multicultural executive and business leader with more than 17 years of experience in IT, strategy and planning areas, with specializations in business studies at Stanford (U.S.A), INSEAD (France), and EGADE (Mexico). He has international experience of working on several projects in other geographical locations including Asia, Europe, the Middle East, and South America, that are focalized in business integration, valuing and acquiring technology-related companies (the Due Diligence process), and integrating IT-related operations to company standards (Mergers and Acquisitions). He also has experience in retail, pharma, services, and manufacturing industries as a partner at an IT services company, in Monterrey, as an External Director in a Big Five consultant firm, and as a IT Planning Director at Tier 1 Retail in Mexico.

Dedicated to Nelson and Enzo, my beloved children, to the woman who supports my dreams, and my mother, who always believes in me.

**Samundar Singh Rathore** is a data warehousing and business intelligence consultant, and a trainer with extensive experience on various data warehousing business intelligence products. He has worked with a large business and analytics organization with specialization in investment and healthcare domain data. He is currently working as a MicroStrategy administrator and architect. He was born in Mundara, Rajasthan, and raised in Mumbai, Maharashtra, which is the financial capital of India. He speaks Hindi and English, and he is also proficient in Java and SQL. He is an active member and the "Guru" of the MicroStrategy official discussion forum at `https://resource.microstrategy.com/Forum/UserProfile.aspx?uid=8557`.

I would like to thank my family and friends for their support and motivation that led me to put my thoughts together for this book. I dedicate this work to my loving mother.

# About the Reviewers

**Chael Christopher** has been working with MicroStrategy since 1999, and writes about MicroStrategy on his blog (www.chaelchristopher.com). In his fifteen years of working in this platform, Chael has worked with clients across a wide range of industries and has seen a diverse set of implementations across financial services, pharmaceuticals, retail, marketing and advertising, supply chain, and higher education. While working with MicroStrategy, Chael has seen the product evolve from a heavy client install (anyone remembers the 16-bit "thunking" layer?) to the Web, and now to Mobile. MicroStrategy continues to innovate across different technology spectrums, and Chael enjoys keeping up with all of the advancements.

**Steven Zagoudis** is the CEO and founder of MetaGovernance Incorporated and Information Governance Technologies Inc. Steven currently holds the VP Marketing Board position for Data Governance Professional Organization (DGPO). He is a Business Intelligence and Information Governance specialist with more than 20 years of experience in extensive business and technical experience. Steven has a proven track record of resolving chronic data infrastructure issues. He has formed data governance functions for numerous organizations. Steven has cross-industry experience in international and domestic organizations, having been based in the U.K. and Germany for several years. He has led the design and implementation of technical and data architectures on complex data warehousing and data mining applications, providing straightforward solutions to business problems. He is a "second surgery" and metadata specialist who refocuses on failed Business Intelligence projects. Steven excels at unlocking information hidden within organizational data, thereby enabling the visibility of trends for improved competitor position. Crain's Business Magazine recognized Steven as a Technology Leader. To Steven, information governance is a journey towards accurate data and meaningful information for business operations, reporting, and disclosure. Steven can be reached at zagoudis@gmail.com.

# www.PacktPub.com

## Support files, eBooks, discount offers and more

You might want to visit www.PacktPub.com for support files and downloads related to your book.

Did you know that Packt offers eBook versions of every book published, with PDF and ePub files available? You can upgrade to the eBook version at www.PacktPub.com and as a print book customer, you are entitled to a discount on the eBook copy. Get in touch with us at service@packtpub.com for more details.

At www.PacktPub.com, you can also read a collection of free technical articles, sign up for a range of free newsletters and receive exclusive discounts and offers on Packt books and eBooks.

http://PacktLib.PacktPub.com

Do you need instant solutions to your IT questions? PacktLib is Packt's online digital book library. Here, you can access, read and search across Packt's entire library of books.

## Why Subscribe?

- Fully searchable across every book published by Packt
- Copy and paste, print and bookmark content
- On demand and accessible via web browser

## Free Access for Packt account holders

If you have an account with Packt at www.PacktPub.com, you can use this to access PacktLib today and view nine entirely free books. Simply use your login credentials for immediate access.

## Instant Updates on New Packt Books

Get notified! Find out when new books are published by following @PacktEnterprise on Twitter, or the *Packt Enterprise* Facebook page.

# Table of Contents

# Preface

This book covers various MicroStrategy 9 capabilities, such as the do-it-yourself approach, import and export of data from or into MicroStrategy, dashboards, scorecards, sharing or distribution of BI reports/dashboards, Cloud BI, and Mobile BI. It also explains standard designs and best practices applicable to each chapter depending on the content and context. This book, from start to finish, will guide you well on the various offerings and practices around creating various analyses using the MicroStrategy platform Cloud service.

## What this book covers

*Chapter 1*, *The Value Proposition of MicroStrategy*, serves as a guide for the self-service concept, for the prerequisites to install and enable the platform, and for preparing a data set used in the book as a reference. At the end of the chapter, the user will understand the self-service concept and confirm that he can apply the DIY approach.

*Chapter 2*, *Mapping Typical Business Needs*, explains to the user the procedure and mechanism, with a very practical approach, of how to map business needs to MicroStrategy components. At the end of this chapter, users will have a clear understanding of how to map their real business to MicroStrategy components.

*Chapter 3*, *Reporting – from Excel to Intelligent Data*, explains the procedures and gives hints to design, build, and deploy BI reports using the MicroStrategy platform. The users will be able to generate their first report and analyze data based on a real business need.

*Chapter 4*, *Scorecards and Dashboards – Information Visualization*, explains the procedures and mechanisms to design, build, and deploy scorecards and dashboards, focalized in the self-service approach. At the end of this chapter, the user will be able to identify the objects and rules of MicroStrategy for enabling dashboards and scorecards. This chapter will also serve as a practical guide for users to develop their own dashboards.

*Chapter 5, Sharing Your BI Reports and Dashboards*, covers the procedures to share reports and the subscription engine. At the end of this chapter, the user will be able to set up MicroStrategy components in order to share the reports to defined audiences in an automatic schema.

*Chapter 6, MicroStrategy and the Cloud*, explains the Cloud offering of MicroStrategy and its usage. At the end of this chapter, the user will be able to identify the Cloud offering of MicroStrategy and see that it is ready to use with basic models.

*Chapter 7, BI Reports at Your Hands*, explains the mobile offering of MicroStrategy, with some practical recommendations and guidelines to enable it. At the end of the chapter, the user will be able to identify the offering of MicroStrategy in the mobile arena, and will be able to design and build a specific model in an iPad and Android device.

*Appendix A, MicroStrategy Express*, explains the various types of MicroStrategy Cloud offerings with relevant screenshots.

*Appendix B, Visualization*, explains various visualization properties and some of the key visualizations in detail.

# What you need for this book

- A MicroStrategy account to enable the Cloud express service
- Microsoft Excel

# Who this book is for

This book is best for a beginner MicroStrategy developer or an analyst who wants to get more insight on various MicroStrategy business intelligence features and offerings. Senior MicroStrategy users can also benefit from the latest product offerings around Cloud and mobile implementations. Also, this book is useful for business users who want to analyze their data on their own using MicroStrategy. Though this book begins with the core concepts of business intelligence with MicroStrategy, it would be helpful to have some prior familiarity on business intelligence products and their applications.

# Conventions

In this book, you will find a number of styles of text that distinguish between different kinds of information. Here are some examples of these styles, and an explanation of their meaning.

Code words in text are shown as follows: "We can include other contexts through the use of the `include` directive."

**New terms** and **important words** are shown in bold. Words that you see on the screen, in menus or dialog boxes for example, appear in the text like this: "Select the **Profit by Employee** filter."

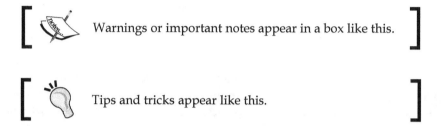

Warnings or important notes appear in a box like this.

Tips and tricks appear like this.

# Reader feedback

Feedback from our readers is always welcome. Let us know what you think about this book—what you liked or may have disliked. Reader feedback is important for us to develop titles that you really get the most out of.

To send us general feedback, simply send an e-mail to feedback@packtpub.com, and mention the book title via the subject of your message.

If there is a topic that you have expertise in and you are interested in either writing or contributing to a book, see our author guide on www.packtpub.com/authors.

# Customer support

Now that you are the proud owner of a Packt book, we have a number of things to help you to get the most from your purchase.

# Downloading the example code

You can download the example code files for all Packt books you have purchased from your account at http://www.packtpub.com. If you purchased this book elsewhere, you can visit http://www.packtpub.com/support and register to have the files e-mailed directly to you.

# Downloading the color images of this book

We also provide you a PDF file that has color images of the screenshots/diagrams used in this book. The color images will help you better understand the changes in the output. You can download this file from: Please use the link: http://www.packtpub.com/sites/default/files/downloads/0044EN_ColorImages.pdf.

# Errata

Although we have taken every care to ensure the accuracy of our content, mistakes do happen. If you find a mistake in one of our books—maybe a mistake in the text or the code—we would be grateful if you would report this to us. By doing so, you can save other readers from frustration and help us improve subsequent versions of this book. If you find any errata, please report them by visiting http://www.packtpub.com/submit-errata, selecting your book, clicking on the **errata submission form** link, and entering the details of your errata. Once your errata are verified, your submission will be accepted and the errata will be uploaded on our website, or added to any list of existing errata, under the Errata section of that title. Any existing errata can be viewed by selecting your title from http://www.packtpub.com/support.

# Piracy

Piracy of copyright material on the Internet is an ongoing problem across all media. At Packt, we take the protection of our copyright and licenses very seriously. If you come across any illegal copies of our works, in any form, on the Internet, please provide us with the location address or website name immediately so that we can pursue a remedy.

Please contact us at copyright@packtpub.com with a link to the suspected pirated material.

We appreciate your help in protecting our authors, and our ability to bring you valuable content.

# Questions

You can contact us at questions@packtpub.com if you are having a problem with any aspect of the book, and we will do our best to address it.

# 1
# The Value Proposition of MicroStrategy

Sometimes when we hear about Business Intelligence platforms and related technology, our first reaction is a web report with tons of data to analyze and review, which has a slow and poor response time. Such platforms require IT-savvy personnel to enable them, as they are complicated to use. However, innovations in information technology have completely transformed the user experience and the way technologies are implemented.

When we implement technologies by ourselves with a do-it-yourself approach, the phenomena is called IT consumerization: technology for the final consumers without the complication of typical IT systems in corporations.

Imagine the same concept and approach translated to the Business Intelligence arena: a BI platform ready to use, simple to install, without specialized training or personnel to enable it, and ready to use our data in order to help us take business decisions.

In this chapter you will learn the scope of the MicroStrategy platform and how to enable it in a do-it-yourself approach without any specialized knowledge or training. Fortunately, MicroStrategy understands this consumerization trend and offers a platform and services that are easy to enable and be used by a business user. In a few hours, a business user is able to analyze information as well as generate reports and dashboards. Also, business users will be able to share their reports in a collaborative schema with their business colleagues, and also on a mobile platform for portability.

# MicroStrategy Business Intelligence suite

The MicroStrategy Business Intelligence suite includes solutions for different business needs and scopes. They can be listed as follows:

- Enterprise reporting
- Scorecards and dashboards
- Advanced and predictive analytics
- Closed-loop BI
- High-performance BI
- Big Data

We will learn more about them in the further sections of this chapter.

## Enterprise reporting

Enterprise reporting supports Business Intelligence by delivering the detailed information that impacts the decision maker's reliance on flexible reporting systems, which present business data in the most consumable format for day-to-day operations. The source for reporting is any transactional system that you use for sales, logistics, and inventory, among others; the opportunities are endless.

For example, reports can be used for production and operational data, invoices, statements and business reports, such as **Profit and Loss (P&L)** statements, and performance and statutory reports.

## Scorecards and dashboards

MicroStrategy provides a powerful platform for scorecards and dashboards that consolidate and arrange numbers and metrics on a single screen.

Scorecards and dashboards may be tailored for a specific role and display metrics targeted towards a single point of view or department.

The Business Intelligence dashboard is often confused with the scorecard. The main difference between the two, traditionally, is that a Business Intelligence dashboard, like the dashboard of a car, indicates the status of the data at a specific point in time and the trending of data. A scorecard, on the other hand, displays progress towards a specific goal over time.

# Chapter 1

# Advanced and predictive analytics

MicroStrategy supports advanced and predictive analysis capabilities, enabling users to perform analyses such as hypothesis testing, churn prediction and customer scoring models within a single interface.

MicroStrategy has built-in support for more than 300 mathematical, **Online Analytical Processing (OLAP)**, financial, and statistical functions.

 A typical example of predictive analytics is using regression techniques to forecast sales, validate profitability, and stay in budget.

# Closed-loop BI

A closed-loop BI is the capability of transforming a BI report's ready-only data into interactive reports that allow the user to capture strategies and data to trigger actions. This provides support to the related data source.

Inventory reports are natural candidates for closed-loop BI; you can change the goal in an inventory report, and based on the metric the system triggers actions to achieve it.

# High-performance BI

High-performance BI is focalized on large amount of data (terabytes, petabytes) without compromising the performance and usability of the reports and dashboards.

Financial risk processes and high-tech applications are the main candidates for these capabilities.

# Big Data

Big Data is about managing large, complex-structured (sales reports) and non-structured (social networks consumer behaviors) data.

 Large retails and end-consumer service providers are the main candidates for Big Data technology.

This book is focused on helping you understand and use the reports and dashboard capabilities of the MicroStrategy platform in Cloud with a do-it-yourself approach.

As a first step, you need to enable your MicroStrategy-platform Cloud service (if you are a MicroStrategy customer, proceed with the enablement using your username and password delivered by MicroStrategy); the detailed instructions and prerequisites are located in *Appendix A, MicroStrategy Express*.

Enabling and using the MicroStrategy Cloud service is as easy as working with the Excel application, and moreover, Cloud applications have very easy and precise steps for performing any task.

The end user interface (menus, options, and so on) in the MicroStrategy Cloud platform (Internet based) will be very familiar to you. Please keep in mind that the final objective is that you will be able to design, build, analyze, and share reports and dashboards to leverage MicroStrategy from your own desktop/mobile device.

**Downloading the example code**

You can download the example code files for all Packt books you have purchased from your account at http://www.packtpub.com. If you purchased this book elsewhere, you can visit http://www.packtpub.com/support and register to have the files e-mailed directly to you.

# Summary

The MicroStrategy portfolio includes several offers for enabling BI solutions. Consumerization trends are now part of the BI arena, and MicroStrategy offers a Cloud version for faster reports and dashboard design without the intervention of IT-savvy personnel; now you as a business manager or director, can do it yourself.

Prior to building the reports, we need to identify the business needs and design the scope of the solution.

# 2
# Mapping Typical Business Needs

What is the best way to start to using the MicroStrategy platform? In this chapter will answer that question.

At first, we need to have our business needs clearly identified in order to design the solution to leverage the platform (the response we are looking for, the location of the data, the issue that we need to solve, the person responsible taking decisions with the information, and so on).

But prior to that, we need to understand what a BI solution is. According to *Forrester Research*:

> *"Business Intelligence is a set of methodologies, processes, architectures, and technologies that transform raw data into meaningful and useful information used to enable more effective strategic, tactical, and operational insights and decision-making"*

When we use this definition, a BI solution also includes technologies such as transactional-source data systems, data transfer process, data repositories, final-user tools, and information distribution and control. But what do all these terms mean to us?

The following list explains the main components of the BI solution:

- **Source data systems**: This is the raw data for our reports. The main source for Business Intelligence data to be analyzed is all data captured, that is, the sales or operational data from the factory, financial transactions, and customer feedback.

- **Data transfer process**: This is the mechanism to extract the raw data into our reports. All of the necessary data must be processed from source data systems to specialized repositories or must be shown to the final users. These data interfaces are called **Extraction, Transformation, and Load (ETL)**.

- **Data repositories**: This is where the processed data of our reports is stored. Depending on the size and the scope of this repository, it could be named data warehouse, where data about the entire organization or most of the organization is stored, or it could be named datamart, where data about individual departments or organizational units is stored.

- **Final-user tools**: They are the end-user applications for the design and generation of reports. Users have special tools that access data warehouses and datamarts, and these tools access the data dictionaries for the document and inform the users about the accessed data and its meaning.

- **Information distribution and control**: It shares our reports via different channels in a secure way. Regular reports, news, and other information must be delivered in a timely and secure fashion through ways such as e-mail, mobile, or web to appropriate personnel.

Now, with these concepts in mind, we need to map our business needs to the BI solution architecture and the MicroStrategy platform in the Cloud or on-premise implementation.

# Understanding the business needs

Let's assume for a moment that you are the commercial director of a retail chain with a massive growth during the last two years, and we need to identify our top five / bottom five stores in terms of revenue, cash flow, and sales representative's rotation, and totalize the profit, operative and net income, and most importantly, determine what is the rationale for identifying the top and bottom performers.

You need this information as soon as possible as your Regional VP is asking for those reports, and you can't wait for your IT fellows or consultants to deliver the reports because your institutional BI platform is under construction or definition, and you do not have any platform in your company for BI or is already in production but requires several processes, documents, testing, and validation in order to meet your requirement.

The business need is already defined, the question to answer is set by your Regional VP, and you have already enabled the MicroStrategy platform in the Cloud; the next step is to define your BI model strategy prior to using the platform.

In the following table, we define — by domain — the business need mapped to the concepts already defined:

| Domain | Strategy |
| --- | --- |
| Business needs (what we need to analyze) | Top five / bottom five stores by revenue, cash flow, and sales rep's rotation, sales and profits summarization |
| Source data systems (where the data come from) | Sales report in Microsoft Excel format |
| Data transfer process (how the data is inserted in our model) | Transfer the Excel file using MicroStrategy import capabilities |
| Data repositories (where the data of our model resides) | Cloud database, enabled during the MicroStrategy setup process |
| Final-user tools (what is the user interface for our reports) | Reports and dashboards design MicroStrategy tools in your own laptop via a web browser |
| Information distribution and control (how the information is shared) | Generate web- and PDF-format reports and send it via e-mail compatible with mobile devices for oblique's access |

The strategy is already defined; the next step is to set up the data model in the MicroStrategy platform. You already enabled the platform according to *Appendix A, MicroStrategy Express*. Besides, you already have your sales report data (that is available for download in order to define the model).

The next step is to load the data and prepare the model.

Preparing our first model by uploading the data in the MicroStrategy platform in the Cloud is a very straightforward process:

1.  Access the platform with your credentials (already enabled in *Appendix A, MicroStrategy Express*).

2.  Select the new dashboard option.

3.  Select your data source: Excel/CSV.

4. Select the `salesrawdatav1.0.xls` file as shown in the following screenshot:

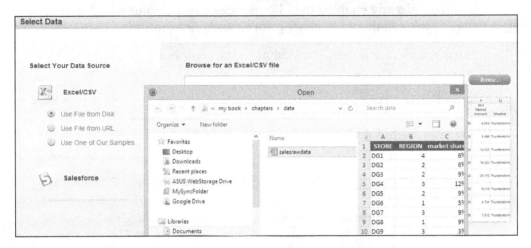

5. Click on **Continue**, and a loading screen will appear as shown in the next screenshot:

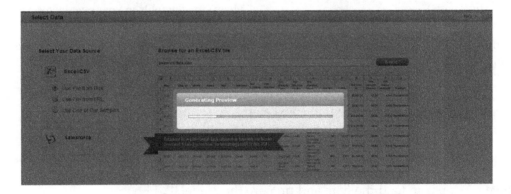

6. The next step is to validate whether the columns of the data are **metrics** (columns used for calculations, cost, revenue, and so on) or **attributes** (columns that organize and group metrics). These metrics and attributes will be used in our reports. Please ensure that only the store, region, market share, and open year columns are attributes when you select the data. Click on the **Continue** button.

> By default, MicroStrategy identifies all numeric columns from the updated data as metrics and non-numeric or semi-numeric columns as attributes; however, you may change the pre-identified column type as per your business.

7. Select the grid option from the **Select a Visualization** dialog (you can change it later to any other type using the change visualization icon from the toolbar).

8. Click on the **Save** button; the system will request the name of the report. Type data load.

The options for data manipulation, reports, and dashboard generation in the platform are the following:

Show and hide datasets (the columns of the Excel file that we upload), filters, navigation options to navigate between pages of the data, edit, insert, and change visualization options, and tools, as shown in the following screenshot, are located at the top-left corner of the main menu of the MicroStrategy Cloud interface:

# Summary

We understood what a BI solution is and its components. We laid out a strategy to solve a business issue that requires a BI platform. We also loaded our data onto the platform to start designing our reports and dashboards without any IT-support personnel or complex software usage, in a self-service, do-it-yourself approach.

It was similar to accessing our preferred social network or the Internet navigation on our smart TV but is now more business-oriented and focalized to solve a business issue and makes better decisions to leverage Cloud-innovative solutions, such as MicroStrategy.

# 3
# Reporting – from Excel to Intelligent Data

Microsoft Excel is the most common tool for business reporting and analytics. Despite its excellent capabilities in that role, the final objective of Microsoft Excel is not concerned with its **Business Intelligence** (**BI**) capabilities. There is a set of specialized platforms and tools for that purpose, and **MicroStrategy** is one of the best providers for BI.

Gartner defines BI as:

> *"Business intelligence (BI) is an umbrella term that includes the applications, infrastructure and tools, and best practices that enable access to and analysis of information to improve and optimize decisions and performance."*

In this chapter, we will learn the reporting concepts for data manipulation and their implementation in the MicroStrategy platform in a simple and practical way. Additionally, we will learn how to use the MicroStrategy platform to its maximum strength. At the end, using MicroStrategy will be as easy as using Microsoft Excel. Please keep in mind that the ultimate goal is to produce BI reports in a do-it-yourself schema without the specialized support from an **IT Guru** or technical driven manuals and instructions.

Let's start with the key concepts that we will use in this chapter and that are needed to start building the BI reports.

The manner in which the information is represented in our reports is crucial in order to generate the desired impact on our BI report audiences. The amount of data, the business question that we need to solve will define the kind of visualization that we use. We will be using **Visual Insight** that is an integral part of MicroStrategy web application and **Cloud** platform service during our demonstrations and exercises.

# Visualization objects – graphs and grids

The following figure shows a graphical representation of information:

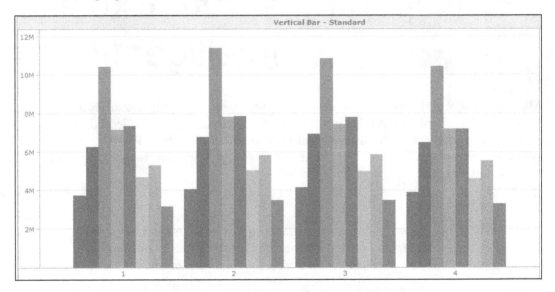

By graphically representing the information in one single view, we can respond to business questions and detect trends and behavior of the information. In this case, we are looking at sales by region, where region **2** is the most profitable.

The following screenshot gives a detailed account of an object that is ideal for data analysis (store in this example) in order to detect specific behavior and review specific values in a grid representation:

| Store | Qrt1 Net Income | Qrt1 Operative Income | Qtr1 Sales | Qtr1 Profit |
|---|---|---|---|---|
| DG1 | 21790.871999999996 | 38229.6 | 449760 | 152918.4 |
| DG101 | 14971.067999999997 | 23763.6 | 452640 | 113160 |
| DG102 | 12329.1 | 21630 | 346080 | 86520 |
| DG110 | 17452.4544 | 29087.424 | 521280 | 161596.8 |
| DG111 | 23706.071999999996 | 36470.88 | 490200 | 151962 |
| DG112 | 12263.545440000002 | 21514.992 | 365280 | 113236.8 |
| DG114 | 11579.04 | 19298.4 | 283800 | 96492 |
| DG115 | 14687.243999999999 | 22595.76 | 342360 | 102708 |
| DG116 | 20765.64 | 35196 | 502800 | 140784 |
| DG117 | 23726.85 | 40215 | 459600 | 160860 |
| DG121 | 8765.528400000001 | 15378.12 | 294600 | 85434 |
| DG122 | 8810.01576 | 15456.168 | 283080 | 73600.8 |
| DG126 | 13301.882880000001 | 20784.192 | 363360 | 94473.6 |
| DG127 | 10180.655999999999 | 16689.6 | 278160 | 69540 |
| DG13 | 14993.510400000001 | 25850.88 | 448800 | 143616 |
| DG131 | 23175.936 | 36212.4 | 517320 | 181062 |
| DG139 | 14729.420159999998 | 25395.552 | 477360 | 133660.8 |
| DG141 | 15324.1308 | 27862.056 | 473040 | 146642.4 |
| DG143 | 18473.49 | 31311 | 417480 | 125244 |
| DG147 | 15269.7006 | 27763.092 | 455880 | 132205.2 |
| DG148 | 16764.1344 | 28903.68 | 463200 | 120432 |
| DG149 | 11626.2216 | 20396.88 | 323760 | 97128 |
| DG153 | 12777.3072 | 22029.84 | 349680 | 122388 |
| DG159 | 19103.182559999997 | 30322.512 | 510480 | 168458.4 |
| DG162 | 14344.711100000008 | 23515.02 | 435480 | 117570.6 |

We can use a combination of graphs and grids in one single view for a quick review of the data without the need to navigate to separate pages; this kind of single view is called a dashboard. So far, the concept is quite similar to Microsoft Excel; however, the main difference is the way in which to produce and design this visualization in MicroStrategy and how to exploit, analyze, and share information with key personnel responsible for making decisions based on the information analysis. But Excel and MicroStrategy are not separate alternatives; in fact, they are complementary and MicroStrategy offers a level of integration from and to the Excel data.

The MicroStrategy platform has the advantage of having more control of the data in a centralized and secure infrastructure instead of the end user's personal computer with tools like Excel. Another key advantage is the management of the official (institutional) formulas and calculations of **Key Performance Indicators (KPI)** for our reports instead of each user generating and defining their own in their local Excel files. More importantly, the ability to share the latest version to the key personnel through different channels such as e-mail, tablets, smart phones, and the Web and making changes in the data, reports, and KPI without affecting the model integrity and assuring all the people have the same version of the data is another feature of MicroStrategy.

In the following table, we will analyze a set of characteristics of Excel and MicroStrategy:

| Characteristics | Excel | MicroStrategy |
| --- | --- | --- |
| Familiar user interface | This is the primary option for power and advanced users (engineers and mathematically oriented users) | This has a simple and easy-to-use interface for business users, managers, and directors |
| Ability to tweak a report | Yes, this provides sorting, filtering, pivoting, or removing a column | Yes, similar to Excel |
| Extensive formula library | Yes | Yes, this accepts formulas for business driven calculation, but do not expect advanced mathematical or engineering formulas |
| Access to multiple data sources | Yes | Yes, this is a must have requirement |
| Ability to "massage" the data | Yes, this is the main function of Excel when reports are needed | No, the data must be changed at the source, not in the BI platform |
| Ideal prototyping environment | No, this is only for single user with lack of flexibility in the end user interface | Yes, this is part of their value proposition: the ability to build models very fast and change it as needed |
| Share/accessing to reports | In Excel, this is limited | MicroStrategy reports can be made available in shared folder location on the Cloud or the on-premise server for review or access by other users |
| Advanced widgets for data representation | Excel supports basic graphs/charts | MicroStrategy has a pool of widgets that is available and easy to use |

For any given requirement, assess if Excel is the best solution or the MicroStrategy platform is better. For example, MicroStrategy allows users to filter, sort, and interact with a report via a web browser and share it to mobile devices. Excel is ideally suited for joining data from multiple data sources for one-off analysis.

# Designing and formatting the reports

The first step is to define the business need that we need to address; we already defined it in *Chapter 2*, *Mapping Typical Business Needs*. The next step is to get familiarized with the MicroStrategy interface. After we log in to the MicroStrategy Cloud platform, we will see a single-page interface, and the main screen that we will use is explained in the following screenshot:

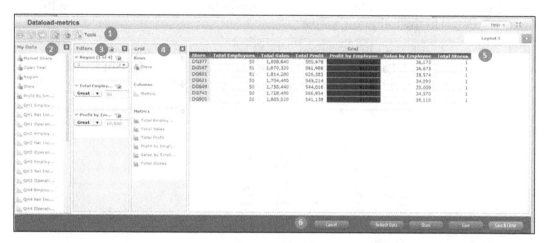

The following are the main options of our menu that we will discuss in this chapter, thus resolving business needs instead of explaining the options one by one:

- The area highlighted as **1** shows the activate or deactivate panel on the top of the window provides us with the **Tools** option

- The area highlighted as **2** shows the **My Data** panel shows the attributes and metrics loaded in our model; it also includes the option to add and modify new metrics

- The area highlighted as **3** shows the **Filters** panel allows us to add and configure filters for data selection, search, and slide

- The area highlighted as **4** shows the **Grid** panel allows us to add rows, columns, metrics, and format our data in a grid or graph view

- The area highlighted as **5** shows the **Grid** section shows the data that is to be analyzed and allows us to arrange it using panels

- The area highlighted as **6** shows the action panel at the bottom allows us to cancel and load data and save our model

[ Use the **Save** button frequently. MicroStrategy Visual Insight does not include auto save, and your work can be lost. ]

Now, it is time to fulfill the first business need: identifying our top five/bottom five stores in terms of revenue (for all the options and menus that we use to design our report; later in this chapter, we will learn the details of the options during this exercise). Perform the following steps:

1.  Proceed with the MicroStrategy Cloud service; sign in and select the **dataload** model already loaded in the platform and the default **Layout** that is created, as shown in the following screenshot:

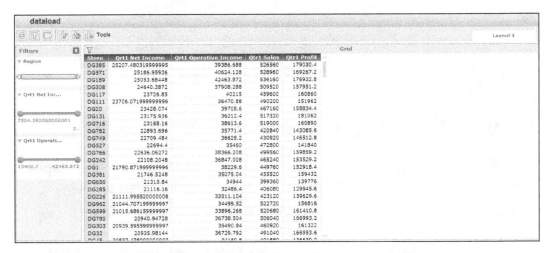

2.  Then, select the filters from the **Filters** panel by clicking on the drop-down button [ 🔲 ] and choose the **Qrt1**, **Qrt2**, **Qrt3**, **Qrt4** options for **Operative Income**, **Net Income**, **Profit**, and **Sales** respectively, as shown in the following screenshot:

3. Now, click on any header of the grid, navigate to **Add to Grid**, and select the same metrics already chosen in the prior step, as shown in the following screenshot:

We already have all the active filters in our main model, and the grid shows all the related information. Now, we need to calculate the total year profit, that is, the result of the sum of **Qrt1 Profit**, **Qrt2 Profit**, **Qrt3 Profit**, and **Qrt4 Profit**. MicroStrategy includes the option of calculated metrics that is needed in our case; the procedure to enable it is as follows:

1.  In the **My Data** panel, select **Insert New Metric…** by clicking on the add button [□] and create a new metric, as shown in the following screenshot:

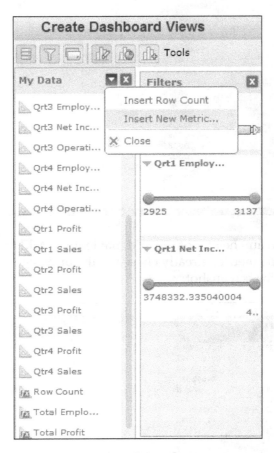

2.  An editor screen will appear; change the name of the metric to `Year Profit` and from the **Available objects** panel, select **Qrt1 Profit**, **Qrt2 Profit**, **Qrt3 Profit**, and **Qrt4 Profit** as shown in the following screenshot:

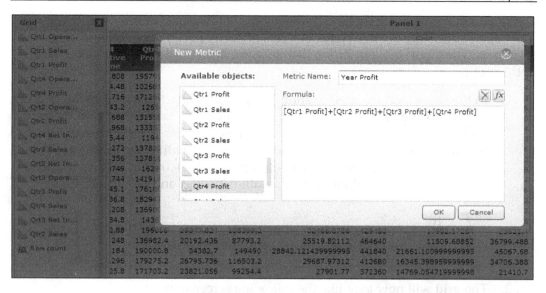

3. Click on **Ok** and the new metrics will appear at the end of the grid; select the **Move Left** option from the metric after the sales metric in the grid.

4. In the **Filters** panel, add the **Row Count** filter; select the **Rank highest** option and the **Qualification** option in **Display Style**. Set it as **Less than** or **Equal to** and then type 5. Next in the grid, navigate to the **Year Profit** column and select the **Sort Descending** option; now, the grid will show the top five performers, as shown in the following screenshot:

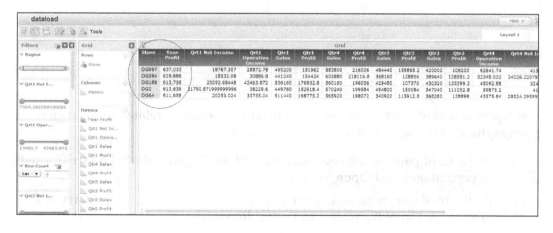

Congratulations! We have already designed our first BI report in less than 10 minutes. This report shows the top five stores in terms of yearly profit; now, we need to format the report. The report is about profit; therefore, we need to add relevant data related to the profit.

1.  In the first place, we need to show the profit by quarter; the best way to do so is via the drag-and-drop option on the **Grid** panel. Select the **Profit** metric by quarter and arrange it beneath the year profit metric. For each quarter profit metric, you can define the number format (click on the metric button  and select **Format**), select the currency for profit metrics as well as the need to arrange and format the sales, operative income, and net income metrics.

> You can use the *Ctrl* key and the mouse for multiple metrics' selection and apply the number format.

2.  The grid will now look like the following screenshot:

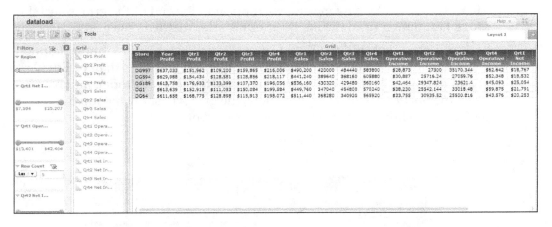

The report is almost ready; now, we need to add information related to the region (geographical zone) and the date when the store opens:

1.  In the **Grid** panel in the rows section, add the **Region, Market Share** (format to percentage), and **Open Year** objects.

2.  As the final step, rename **Layout 1** to `Top 5 Report` (click on the **Layout 1** label at the top of the screen and select **Rename**) and save the report.

We already know the most profitable store: **DG997** from **Region 1**, opened in **2001** with **Market Share** of **12%**. In order to visualize the complete information, the next step is to generate a graph based on our first report. The steps are as follows:

1.  Copy the report to **New Panel** as shown in the following screenshot:

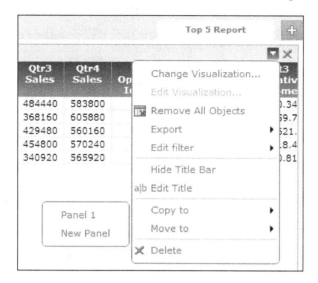

2.  Switch to **New Panel** and select **Change Visualization…**.
3.  Select a bubble (metrics on y – x axis) as shown in the following screenshot:

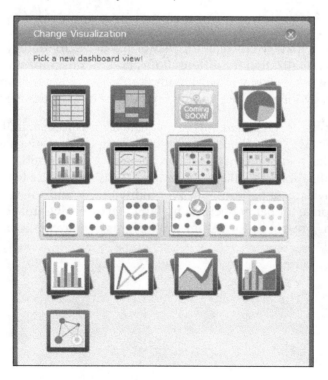

4. In the **Graph Matrix** panel in the **Y Axis** section, add a new formula: **Total sales**, that is, the sum of **Qrt1** sales to **Qrt4** sales.

5. Rename the panel (click on the drop-down button [▼] on the top and select **Rename)** and type First Graphics.

6. Save the panel. The resulting graphic is similar to the following screenshot:

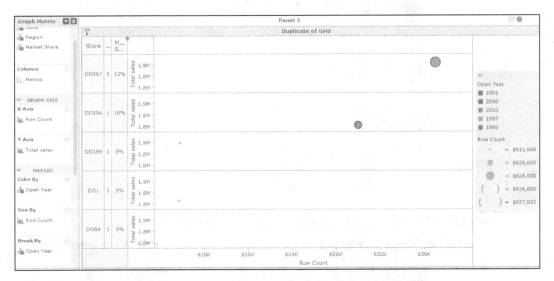

This the first graph that we create in the platform, but MicroStrategy offers wide options for data visualization for different purposes. Some graphs require specific data in order to work. When you work with graphics, a new panel appears in the main MicroStrategy interface, depending on the type of graphics we are using.

In *Chapter 4, Scorecards and Dashboards – Information Visualization,* in the visualization objects section, we will use all the graphics options and alternatives for dashboard creation. Please keep in mind that the MicroStrategy graphs option provides interactive visualization capabilities that enable decision makers to dynamically explore ideas, investigate patterns, uncover hidden facts, and share those insights across the enterprise for better decision making. An extensive suite of customizable graphical options presents information and insights that are not easily detected in grid formats.

The main features of MicroStrategy do-it-yourself schema are as follows:

- Provides self-service access to dynamic, visualization environments
- Reduces the overdependence on IT for ad-hoc requests
- Enables sharing of information via dashboards

# Analyze data

Besides the grid and graphical objects, a key component of the MicroStrategy platform is the ability to analyze and drill the information in order to resolve business issues, detect trends, opportunities or risks, as well as to define different views of data for different audiences. In this section, we will learn about these capabilities and use them.

# Advanced metrics and attributes

MicroStrategy has the capability to add your own metrics (columns used for calculations, cost, revenue, and so on) or attributes (columns that organize and group metrics) besides the data loaded in the model. For advanced analysis, attributes and metrics can leverage mathematical and financial formulas; for example, it is possible to calculate average, standard deviation, sum, and variance as a new attribute, and aggregation and advanced mathematical formulas as the new metrics.

Let's start with some definitions and platform usability. The main interface includes three panels for metrics' and attributes' definition: **My Data** where the attributes button [⬛] and metrics button [⬛] are managed, **Filters** where the attributes and metrics can be used for data filtering, and **Grid** where we can adjust the grid-managing rows, columns, and metrics of our data. These three main panels are managed by clicking on the drop-down button [⬛] for options.

The drop-down button [▣] appears above the options of these panels, and the related options are hidden by default.

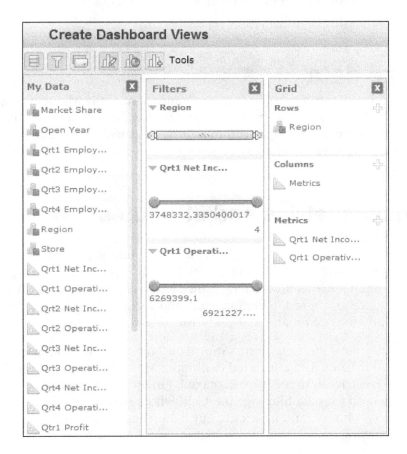

In order to start using this option, please proceed to generate a new model with the same raw data provided and used in *Chapter 2, Mapping Typical Business Needs*, and leave it with the default options. But mark the **Employees** field as a metric, select the **Grid** option, and name the model `dataload-metrics`. The model will be similar to the following screenshot:

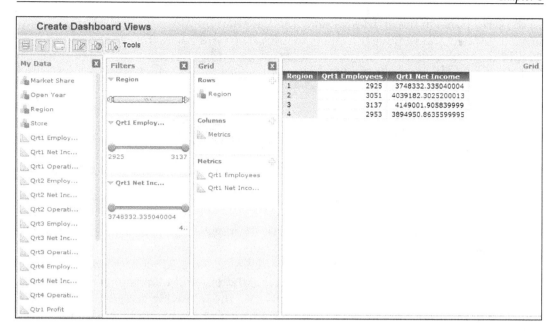

Generate **New Metric** named `Total Employees` from the **My Data** panel option and edit the formula with the sum of **Qrt1 Employees** to **Qrt4 Employees**, as shown in the following screenshot:

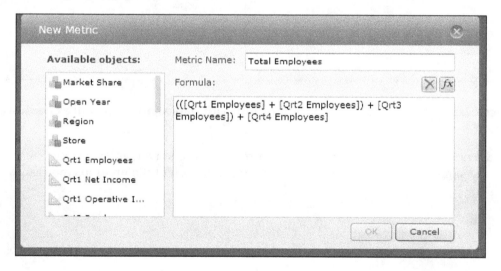

In the **Grid** panel in the metrics section, add the **Total Employees** metric that we already created and remove the others metrics. The **Grid** panel will show only the **Total Employees** metric by region. Repeat the procedure; but now, add a metric for **Total Sales** and **Total Profit**. The **Grid** panel is now similar to the following screenshot:

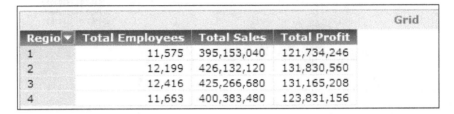

| | | | Grid |
|---|---|---|---|
| Regio ▼ | Total Employees | Total Sales | Total Profit |
| 1 | 11,575 | 395,153,040 | 121,734,246 |
| 2 | 12,199 | 426,132,120 | 131,830,560 |
| 3 | 12,416 | 425,266,680 | 131,165,208 |
| 4 | 11,663 | 400,383,480 | 123,831,156 |

Now, let's calculate the revenue and profit by employee. This is a new calculated metric that is the result of dividing the total sales/total employees and the total profit by the total employees. We need to create two new calculated metrics from the **My Data** panel by selecting the total employees, total sales, and total profits metrics that we created earlier to create new ones as shown in the following screenshot. Also, in the **Grid** panel in the metrics section, add the new metrics that are already created.

MicroStrategy offers a broad scope of formulas for data analysis as well as the **Row Count** metric for data aggregation.

Add a new metric using the row count formula and rename it as Total Stores; use the same procedure that we used for the generation of the previous metrics.

This kind of metric is very useful to identify how many rows (in our case, **Store**) are aggregated in the attribute (in our case, **Region**). Now, the **Grid** panel shows the results similar to the following screenshot:

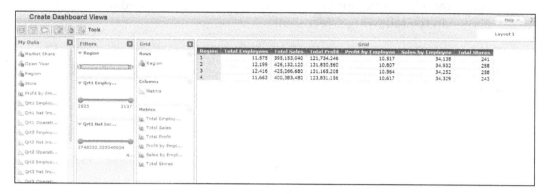

We already transformed our raw data in the BI report in a very simple way without technical complexity.

We were able to determine that **Region 2** is more profitable with fewer employees than **Region 3** with the same number of stores. This kind of information triggers business actions to improve results such as improve sales clerk training and customer centric approach or avoid the rotation of employees in **Region 3**.

# Thresholds

The MicroStrategy platform includes threshold functionality for the metrics in our models in order to detect the relevant data with a visual indicator based on the business rules that we define. For example, let's define the following business rule for the **Profit by Employee** metric:

- If the profit by employee is higher than 10,800, the status is green
- If the profit by employee is higher than 10,600 but lower than 10,799, the status is yellow
- If the profit by employee is lower than 10,600, the status is red

In order to activate the threshold, perform the following steps:

1.  In the **Grid** panel, select the **Profit by Employee** metric in the metric section.

2.  Select the **Threshold** option and eliminate the two intermediate ranges, as shown in the following screenshot:

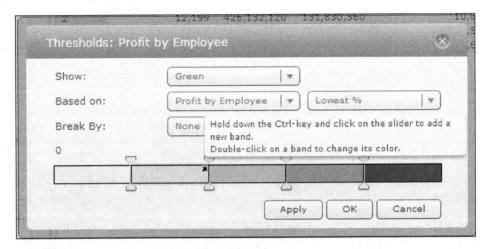

3.  Select the first range from the left, double-click and select the color on the left, and move the slide to the value 10,600.

4.  Select the second range from the left, double-click and select the color on the right, and move the slide to the value 10,799.

5.  Click on the **OK** button.

6.  The threshold is similar to the following screenshot:

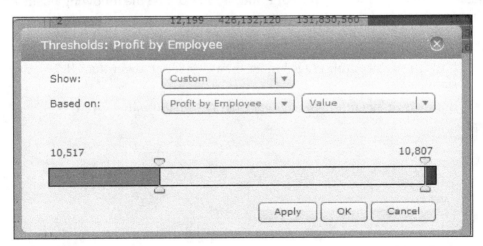

The **Profit by Employee** metric in the **Grid** panel changes color according to the business rule already set in the threshold, as shown in the following screenshot:

| Region | Total Employees | Total Sales | Total Profit | Profit by Employee | Sales by Employee | Total Stores |
|--------|-----------------|-------------|--------------|--------------------|--------------------|--------------|
| 1 | 11,575 | 395,153,040 | 121,734,246 | 10,517 | 34,138 | 241 |
| 2 | 12,199 | 426,132,120 | 131,830,560 | 10,807 | 34,932 | 258 |
| 3 | 12,416 | 425,266,680 | 131,165,208 | 10,564 | 34,252 | 258 |
| 4 | 11,663 | 400,383,480 | 123,831,156 | 10,617 | 34,329 | 243 |

Threshold functionality is a powerful characteristic of MicroStrategy. It helps managers and directors to visualize what is good or bad in one shot and focalize the analysis according to their business needs.

In the previous example, **Region 2** is the top performer in terms of profit by employee and **Region 1** and **Region 3** are the lower ones; in particular, **Region 1** due the low sales by employee in that region.

# Drill down

We already created a model with calculated metrics, and we detected some information that was relevant for the business at high level. Now it is time to analyze and manipulate data leverage of the MicroStrategy platform in order to discover what is behind the data. Let's start with the drilling capability.

First, we need a common definition of what it means to drill. Drilling is a capability that takes the user from a general view of data to a specific one. For example, our report that shows the sales revenue by region allows the user to select a store, click on it, and see what the market share of this store is. He can click on it again and see when the store opens. It is called drill down because it is a feature that allows you to go deeper into the more specific layers of data or information that is being analyzed. The level of the drill depends on the definition of attributes in our model; in our case, **Region**, **Store**, **Market Share**, and **Open Year** are the attributes.

## Benefits of drill down

The benefits of drill down are as follows:

- **Gain instant knowledge of different depths of data**: In mere seconds, drill down answers questions such as which regions of my national sales figure are performing better, which stores are underperforming, and which store is driving revenue within each region.
- **See data from different points of view**: Drill down allows us to analyze the same data through different reports, analyze it with different features, and even display it through different visualization methods.

- **Keep reporting load light and enhance reporting performance**: By presenting only one layer of data at a time, features such as drill down lighten the load on the platform at query time and greatly enhance the reporting performance.

The best place to start to drill down our model is in the main grid view. Click on the drop-down button [⊡] in the **Region 1** row and select **Store**. The **Grid** panel now shows the stores that belong to **Region 1** with the metrics that we already defined in the grid. With the threshold activated in the **Profit by Employee** metric, another click on **Store DG1** to drill **Open Year** will show the year of opening of the store, as shown in the following screenshot:

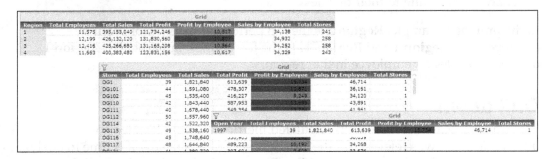

After various drills, you will need to go back to the initial view again; to start with other analysis, click on the filter button [▽] located at the top left of the grid.

Another alternative to drill the information is to add a new row in the grid. This avoids one drill step (other query to the model), but the grid information is loaded with more data and the response time will be compromised.

In order to activate this alternative in the **Grid** panel, add a new row in the **Rows** section and select **Open Year**. Now, grid the **Open Year** group of the store by **Region**, as shown in the following screenshot:

| | | | | | Grid |
|---|---|---|---|---|---|
| Region | Open Year | Total Employees | Total Sales | Total Profit | Profit by Employee |
| 1 | 1990 | 774 | 25,807,320 | 7,916,023 | 10,227 |
| 1 | 1991 | 577 | 20,023,440 | 6,170,646 | 10,694 |
| 1 | 1992 | 424 | 15,300,720 | 4,633,124 | 10,927 |
| 1 | 1993 | 375 | 12,825,600 | 4,004,905 | 10,680 |
| 1 | 1994 | 467 | 15,749,880 | 4,755,832 | 10,184 |
| 1 | 1995 | 425 | 14,289,720 | 4,380,696 | 10,308 |
| 1 | 1996 | 494 | 16,389,960 | 4,990,045 | 10,101 |
| 1 | 1997 | 781 | 28,076,880 | 8,875,760 | 11,365 |
| 1 | 1998 | 429 | 14,430,120 | 4,513,558 | 10,521 |
| 1 | 1999 | 373 | 12,892,320 | 3,985,936 | 10,686 |
| 1 | 2000 | 438 | 15,608,160 | 4,787,762 | 10,931 |

# Filters and slices

Filters and slices are a key functionality of BI. Filters are used in reporting and analysis, for example, to restrict data to a certain region of stores, certain product groups, or certain time periods.

Another example is that you want to reevaluate the transaction data in your model by a factor of 10 percent. However, you only want to perform the reevaluation for certain groups of customers. In order to do this, you create a filter that contains the group of stores for which you want to reevaluate the data. MicroStrategy, offers the filters and slices functionality using attributes and metrics. This functionality can be enabled in the **Filters** panel of the main screen of MicroStrategy and you can add more filters whenever you need them. Add a new filter by clicking on the drop-down button [🔲] in the **Filters** panel using the **Total Employees** metric, and be sure that the region filter is enabled and remove the other filters.

For each filter, it is possible to set different properties depending on whether the filter is based on an attribute or metric:

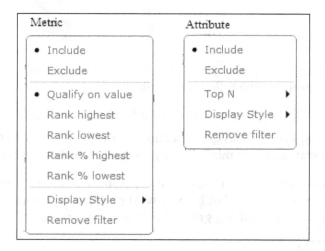

The **Display Style** option allows you to select a **Slider** or **Qualification** option (in case of metrics) and **Check boxes, Search box, Slider, Radio buttons,** or a **Drop-down** list (in case of attributes). For the **Region** filter that has already been created, select the drop-down menu, and for the **Total Employees** filter, select the **Qualification** option. Now, let's assume that we need to analyze **Region 3** in detail. It consists only of stores with more than 50 employees in order to detect issues in the **Profit by Employee** column according to the threshold defined previously. In the **Region** filter, select **Region 3** from the list, and select **Greater than** or **Equal to** and type 50
in the **Total Employees** filter.

> Please be sure to remove the **Open Year** row from the **Rows** section in the **Grid** panel in order to remove that consolidation.

The **Grid** and **Filters** panels looks like the following screenshot:

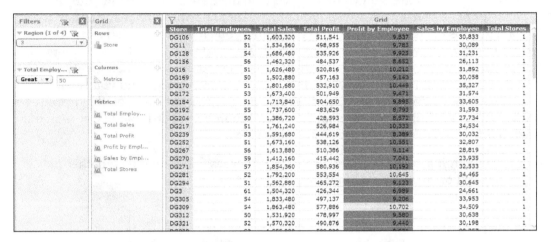

Now, in order to detect why only few stores are in a particular color in the **Profit by Employee** metric, perform the following steps:

1. Add a new filter by selecting the **Profit by Employee** metric.
2. Change **Display Style** to **Qualification** and type 10,800 (this is our value for the green status in that metric); the grid now shows only seven stores.

This information is relevant from the business perspective because **Store** is the benchmark in **Region 3**, which that is the lowest performing region in terms of **Profit by Employee**. The resulting grid is the following screenshot:

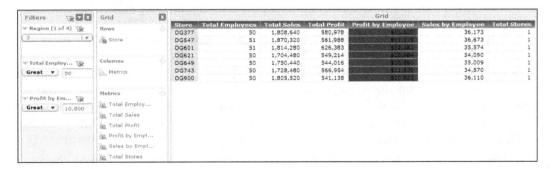

# Advanced options for filters

The filter capabilities in MicroStrategy allow us to group, pivot, and find specific values for detailed analysis. These advanced options are located in the **Filters** panel, and the options are enabled by the type of filter: attribute or metric.

# Finding the values

If we need to find a specific value in our data, it is possible to do that with the **Region** attribute or the **Total Sales** metric. Assume that we need to find the performance of three specific stores with some issues: **DG626**, **DG828**, and **DG212**.

In order to do so, please add a new filter in the **Filters** panel using the store metric, and select **Search box** in the **Display Style** option. Then, type the store's ID: DG626, DG828, and DG212. Now, the **Grid** panel will show the specific data of those stores, as shown in the following screenshot:

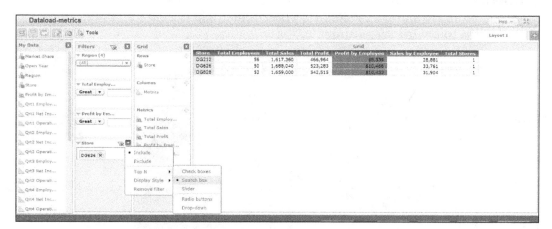

 The size of the box for typing the values that are to be found is limited. Use the scroll ball provided in order to validate your entries or resize the filter using the mouse.

For calculated metrics similar to **Total Employees** or **Total Profit**, it is not possible to search for direct values, but MicroStrategy offers the following specific operators in order to search for specific data:

- Equals
- Does not equal
- Greater than

- Greater than or equal to
- Less than
- Less than or equal to
- Between
- Not between
- Is null
- Is not null
- In
- Not in

If we select equals and the specific data in the options, the query will work like a direct search alternative. For example, let's search for stores with a total of 50 employees only. In the **Total Employees** filter, select the **Display Type** option as **Qualification**, select **Equals**, and type 50 (please be sure to clear other filters, and select the **Store** attribute in the **Rows** section in the **Grid** panel); the results are shown in the following screenshot:

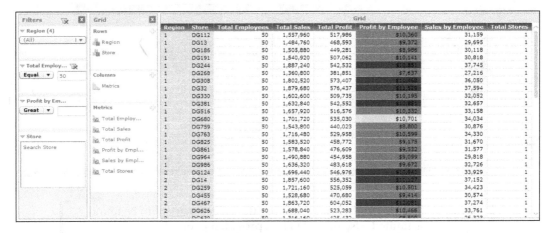

The grid shows all the stores with **Total Employees** equal to **50**; in fact you can combine several filters with search type in order to solve more complex queries. The other benefit is that the entire **Filters** panel works together, regardless of the type of filter; for example, based on the previous result, search for stores with **Total Employees** equal to **50** only from **Region 3**. The **Region** filter is already activated; just select the store from the list in **Region 3**. Now, the grid shows only the stores with **50** employees within **Region 3**, and we can continue to combine filters. We already selected **Region** and **Total Employees**; now, select the **Greater than** option in **Profit by Employee** and type 10,800.

The grid now shows only the stores from **Region 3** with 50 employees and profit by employee greater than 10,800, as shown in the following screenshot:

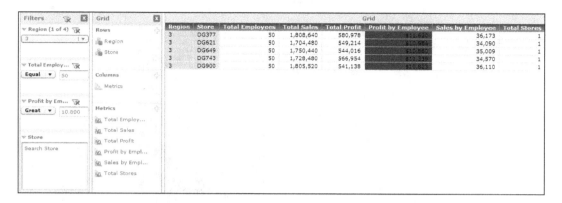

 If we want to reset the filters without deleting them, click on the clear filter button [▓] on the right-hand side of the filter for further queries.

# Including and excluding data

Each filter has the option to include and exclude data, depending on the filter (only works for filters based in metrics) and the display style. The **Include** option in the filter considers the information in the filter for the query and the **Exclude** option does not. It is the best way to search for a specific group of data that belong or do not belong to a specific criteria; for example, if we want to search for stores that were opened in the year 2000 and then search for stores that weren't open in the year 2000 (for benchmark purposes), we need to perform the following procedure:

1. Add a new metric filter, Open Year.

2. Change **Display Style** to **Search box**.

3. Make sure that the **Include** option is selected.

4. Type 2000 in the filter that has already been created.

5.  Be sure that the **Rows** section is added in the **Grid** option panel and the **Store** metrics. The configuration is shown in the following screenshot:

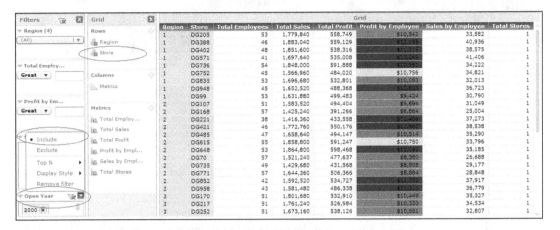

Now, we need to execute the same query but with the stores whose open date consists of all the years except 2000. In the **Open Year** filter option select the **Exclude** option; the **Grid** panel now shows all the stores except those whose open date is in the year 2000. This functionality allows us to execute reports with information filters in a very simple and straightforward process without the complexity of defining database queries or other IT programming languages. Remember, the approach is a do-it-yourself schema.

## Showing the total value of the results

Now, you might be wondering how to totalize the results of your filters and queries in the result grid. In order to have the facts and results of your reports, MicroStrategy offers the possibility to show the totals of your metrics with different formulas such as **Total**, **Average**, **Maximum**, **Minimum**, **Count**, **Geometric Mean**, **Median**, **Mode**, **Product**, **Standard Deviation**, and **Variance**. This functionality is enabled in the results **Grid** panel by clicking on the drop-down button [▣] of the first metric header ,in this case **Region**, in the results **Grid** panel and selecting the **Show Totals** option that you need for your report, as shown in the following screenshot:

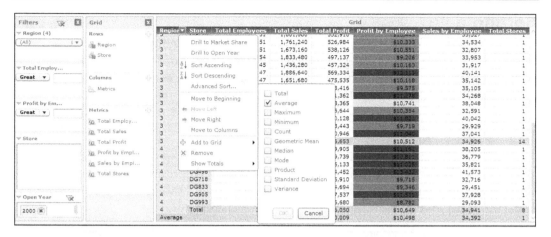

Select the **Total** and **Average** options and run the filter of **Open Year** that contains the value 2000 with the **Include** and **Exclude** options again. At the end of the **Grid** panel (scroll down) you will see the total. The following screenshot shows a total of **43** stores opened in year 2000:

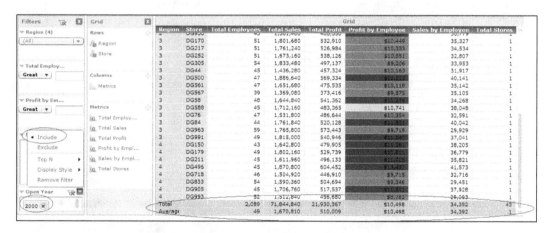

 The totals that you select will remain in your results grid; even if you change the filter and add or remove columns and rows in the grid, the totals row is not removed.

## Sorting data

We already filtered and totalized information in our grid. What if we need to sort the stores by **Profit by Employee** and **Total Employees** in order to detect the top ten? MicroStrategy offers the functionality to sort the grid results. In order to enable it, click on the drop-down button [▣] in the required attribute or metric section to sort the results. In our case, select **Profit by Employee** and sort by clicking on the **Descending** option; the grid will arrange the results sorted by **Profit by Employee**.

It is possible to combine various metrics or attributes in the sorting procedure. In order to enable it, select **Advance Sort** and the required columns and rows to be combined. In our case, select the **Descending** option for both **Profit by Employee** and **Total Employees** as shown in the following screenshot:

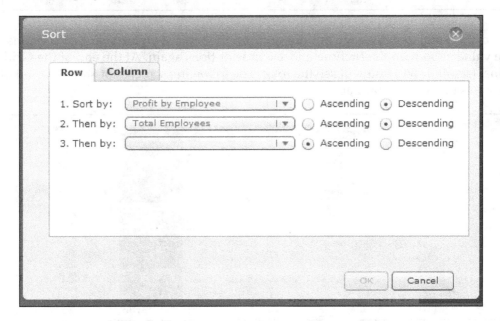

The results in the grid are sorted by the defined rule as shown in the following screenshot:

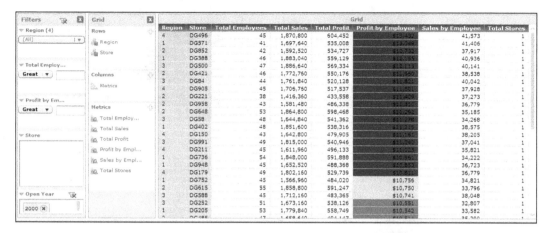

# Ranking

Ranking allows you to choose the rank level at which to return the results of the report. For example, our report contains the **Total Market Share** attribute and the **Profit by Employee** metric that you want to filter so that you can see only the top or bottom 10 stores of **Profit by Employee**.

Instead of generating a filter, sort the results and look for the data as we previously did. The ranking option allows us to generate the data in simple steps to filter it, based on metric value, rank, or rank percentage:

1.  In the **Filters** panel options, reset all the filters selected in order to start a new query.

2.  Select the **Profit by Employee** filter.

3.  Click on the drop-down button [▣] and select **Rank highest** for top performers.

4.  In the filters section, type 10 and in the lists section, select the **Less than** or **Equal to** option.

5. Apply the filter. Now, you will see the top 10 stores in a single data grid in a single step, as shown in the following screenshot:

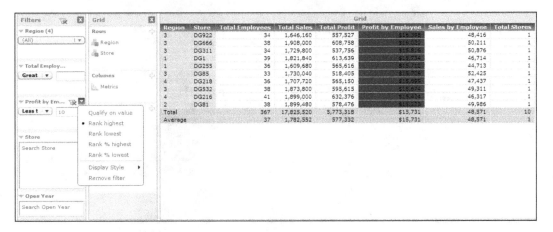

# Summary

In this chapter, we learned the required functionality of MicroStrategy in order to build our BI reports for data analysis and transform data into valuable information for our business needs.

We learned the basic concepts of BI reports, attributes, and metrics. We defined the calculated metrics that leverage the formula options for specific calculations. Also, we learned how to filter data, find specific information, use ranking and threshold functionality, and manage the grid by sorting and summarizing information without complex software installation and IT or BI guru support. We are now able to generate our own reports and use the MicroStrategy platform for our specific reporting needs. Also, we can generate a basic graph for visual representation. The next chapter covers data visualization, and dashboard design and generation in detail.

From the value proposition stand point, we defined the differences between Microsoft Excel and MicroStrategy platform. As you may agree now, MicroStrategy is the best among the breed platforms for BI reports in the market, and Microsoft Excel is the best for data calculations and more sophisticated reports in term of calculations. In the next chapter, you will see more differences between the two.

# 4
# Scorecards and Dashboards – Information Visualization

The concept of scorecards and dashboards has become increasingly popular as organizations discovered their ability to communicate complex information. The terms are sometimes used interchangeably, but there are important distinctions between a scorecard and a dashboard.

While both scorecards and dashboards display performance information, a scorecard is a more prescriptive format; a true scorecard usually includes components such as perspectives (groupings of high-level strategic areas), objectives (verb-noun phrases pulled from a strategic plan), measures (attributes or metrics in our case), and stoplight indicators (red, yellow, or green status, which is managed by the thresholds in MicroStrategy).

Most dashboards are simply a series of graphs, charts, gauges, or other visual indicators that a user has chosen to monitor, some of which may or may not be strategically important.

 Traditionally, the main difference between the two is that a Business Intelligence dashboard, such as the dashboard of a car, indicates the status at a specific point in time. On the other hand, a scorecard displays progress over time towards a specific goal.

In this chapter, we will learn how to design a scorecard and a dashboard from the business and MicroStrategy platform perspective in a do-it-yourself schema. It has the following benefits:

- **Builds your own dashboard quickly**: Using prebuilt templates for both web and mobile, you are able to create publishing-quality, highly-visual dashboards quickly and effortlessly without needing any IT support

- **Quickly identifies trends and outliers**: MicroStrategy offers dozens of visualizations to choose from; and with the drag-and-drop functionality, changes can be made to the data until we find what is needed

- **Allows access to data set**: MicroStrategy enables business users to quickly and easily analyze data in any format, size, or location

- **Explore data to support a hunch**: When you develop your own dashboards, you are able to focus specifically on the data that you want to see

# Scorecards versus dashboards

The definition of a scorecard and a dashboard is now clear, but how do they look? Where do we start? When should we use a scorecard or a dashboard? There is no perfect answer to these questions; however, the rest of this chapter offers a practical guide to define the best option for your business needs.

A scorecard typically looks like the following screenshot:

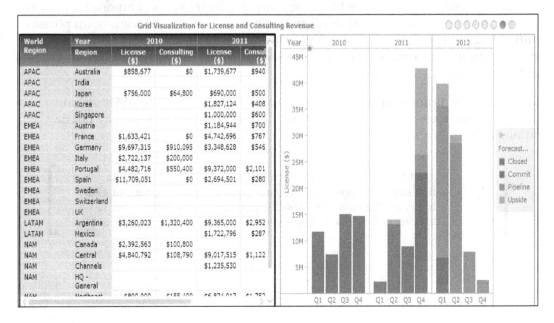

A scorecard displays progress towards specific goals and performance measurements over time. It also validates the measures in order to take actions for corrections, that is, it outlines high-level grouping of data, such as region, year, and quarter using licensing and consulting revenue as a metric, along with the threshold to highlight spotlights.

A dashboard looks like the following screenshot:

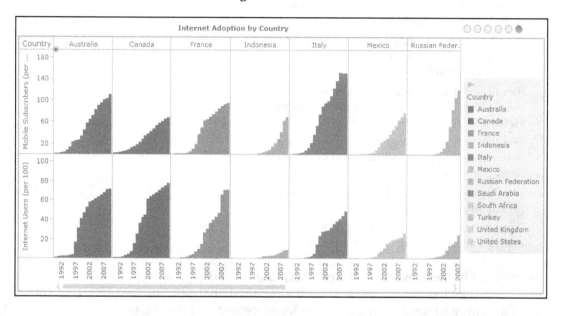

This indicates the status of the analyzed data at a specific point in time. The best way to start designing is by identifying your specific business need.

The following table typifies the more common types of scorecards or dashboards:

|  | Operational | Tactical | Strategic |
| --- | --- | --- | --- |
| Application emphasis | Monitors operations | Reviews progress | Measures performance |
| Users | Supervisors | Office workers | Executives and management |
| Scope | Operational | Individuals | Top management |
| Information | Detailed | Detailed summary | Summary |
| Updates | Intra day | Daily/weekly | Monthly/quarterly |
| Looks like | Dashboard | Dashboard/scorecard | Scorecard |

As you can see, dashboards and scorecards have different scopes. The other guide that we can use to select what kind of visualization to use is explained in the next table in a more practical way:

|  | Scorecard | Dashboard |
|---|---|---|
| Requires a simple snapshot of the information to show the status to the executive audiences, with drill-down capability for validating the data |  | X |
| Requires detailed analysis of the metrics via drill-down to understand the rational of the data or to detect trends | X |  |
| Requires linkage of strategic, tactical, and operational indicators in order to validate their status and performance | X |  |
| Focalized on top-management audiences for an easy use and understanding of the information in a short period of time |  | X |
| Need to show stoplight indicators (red, yellow, green), managed by specific business rules | X |  |
| Appropriate for advanced users in terms of data analysis and exploring data to support a hunch | X |  |

As outlined in the previous table, one may choose to design a dashboard or scorecard, whichever suits the best for the required analysis. The next step is to get familiar with various design guidelines for scorecard and dashboard.

# How to design a scorecard or dashboard

With a clearly defined business need and the basic guidelines for selecting a scorecard or dashboard in place, we are ready to start building our first visualization report to leverage all of the benefits of the MicroStrategy platform. But it is not just about creating a scorecard or dashboard; we need to create an effective scorecard or dashboard. Please consider the following best practices when you start designing your solution:

- From the functional perspective:
    - Provide a high-level overview of the information
    - Provide visual indicators to alert users about important information

- Provide interactivity and personalized information so that an individual can easily understand the data, and has just the information they need to do their job or monitor performance

- Provide guided-analysis navigation of the constrained amount of information

- From the usability perspective:

  - Interactive visualization of key metrics

  - Intuitive user interface and navigation

  - Ability to manage and monitor metrics effortlessly

  - Drill down or through for root cause analysis

From the technological standpoint, design the scorecard or dashboard layout based on how you need to view the data, and leverage the best practices already described.

Please keep in mind that scorecards or dashboards should not do the following:

- Provide advanced analysis capabilities, since that is done by an analysis tool

- Show large amounts of detailed information, since that is a report

- Provide access to ad hoc and open-ended information, since that is done by an ad-hoc-analysis tool

 We will re-use the models that we already created in the previous chapters. We have already defined the key metrics and data grids that are needed to build our scorecards and dashboards.

Now, with all of the definitions clear and some ground rules in place in order to help you define when to use a scorecard or dashboard and how to implement it, the next step is to familiarize yourself with the MicroStrategy platform so that we can start building the first visualization.

# MicroStrategy interface – panels, layouts, and visualizations

Let's start with an explanation about the MicroStrategy interface for the generation of scorecards and dashboards. Log in to MicroStrategy Cloud Express and select the **Dataload-metrics** model.

The first concepts that we need to understand are the layouts, panels, and visualizations:

- **Layout**: This is a new sheet in our model to leverage the same data; if the model is a book, the layout will be a chapter
- **Panel**: This is a subset of a visualization object (grid, graphics) within the layouts; the panel will be a section within the chapter of the book
- **Visualization**: This is the representation of the data through a grid of graph; the visualization will be like the content of the section of the book

Please review the following screenshot for the graphical presentation:

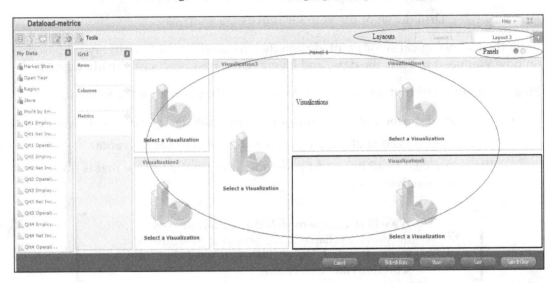

The preceding screenshot is a newly-created empty layout. Each box is saying **Select a Visualization**, and they will be used to select an appropriate visualization from the available visualizations.

In order to generate these objects for our visualization reports, please follow the ensuing procedure:

- Create a new layout by clicking on the add layout button [ ＋ ] located at the top-right corner of the screen, next to the default layout
- Create a new panel from the **Tools** menu located in the main options at the top-left corner of the main screen
- Create five visualizations via the insert visualization button [ ⬚ ] shown in the preceding screenshot

We will re-use the models that we already created in the previous chapters. We have already defined the key metrics and data grids that are needed to build our scorecards and dashboards.

Each object has its own features for specific functionality that is activated by clicking on the drop-down button [].

The features of the objects will help us in the definition of the scorecards and dashboards. The more relevant and frequently used ones are listed in the following table.

# Features

The following table describes the features of the components that will help us in the definition of the scorecards and dashboards:

| Features | Description |
| --- | --- |
| Layouts<br>• Rename the title | For self-explained scorecards or dashboards, this function allows us to define a title of the layout of the report. |
| Panels<br>• Add panel | Add new panels for more visualizations within reports, this option will help us generate more visualizations in the same reports. It is possible to add more than 20 panels by layouts. |
| Visualizations<br>• Edit visualization | With this option, it is possible to modify visualizations (change from one type to another, that is, the graph type). |
| • Export | Export the visualization data or image to Excel or PDF. |
| • Use as a filter<br>• Hide title bar<br>• Edit title | Use the visualization objects as filters in order to interact with the data; if one visualization changes the other visualizations are affected. |
| • Copy to<br>• Move to | Copy and move visualizations to other panels within the same layout. |

A key feature that is often used is exporting the layout and the panel that we created as an image. This is done in order to share a screenshot of the data or include it in a presentation or other document.

This option is enabled via the **Tools** menu by selecting the **Export As Image** option, as the shown in the following screenshot:

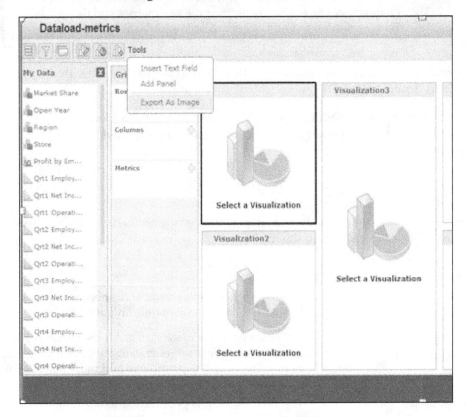

MicroStrategy exports an image (in a .png format) with the data of the report, along with the options required to use the report; but the options for editing it are not included. When we create a panel with more than two visualization objects, it is possible to maximize a specific visualization for a clear view and review. This is shown in the following screenshot:

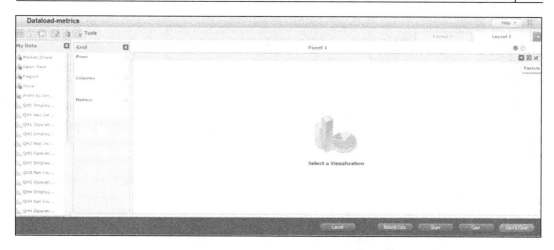

We restore a panel as shown in the following screenshot:

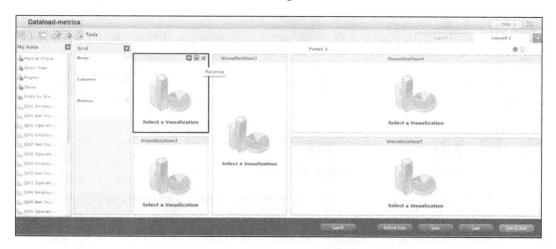

Also, for a more clean and lean report, each visualization object is capable of hiding the title bar (click on the drop-down button [ 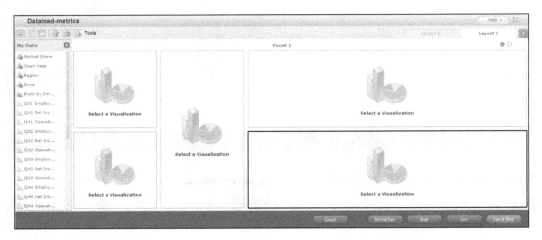 ] and select **Hide Title Bar**). The following screenshot shows our report without the title bar of each visualization object:

When we design our scorecard or dashboard, besides the business needs analysis, we need to consider how users consume the visual information in order to increase their use and adoption.

## Usability best practices

We need to consider the following usability best practices:

- **Reading style**: This is usually determined by how a person reads (left to right or top to bottom)
- **First place to look**: The top-left section is likely to be the first place a user will look
- **Last place to look**: The bottom-right section will likely be the last place a user will look
- **User attention**: Larger graphical items will grab a user's attention anywhere on the dashboard

- **Plan out our dashboards and scorecards**: This means it should fit on a 1024 x 768 single screen

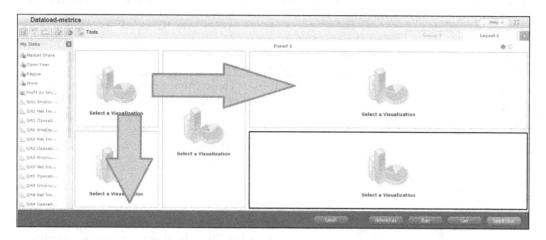

# Visualization objects

MicroStrategy offers several visualization objects for the creation of our scorecards and dashboards. Each object can be used in our report, and it is used to show and support our decision. But, what is a visualization object? Visualization is a visual representation of the data in MicroStrategy, such as a grid or a graph.

You can add visualizations to an analysis to provide multiple ways for a user to display and interact with the data in the analysis.

# Visualization types

An analysis can contain many visualizations. In the following table we describe each object and its primary function:

| Visualizations | Description |
|---|---|
| 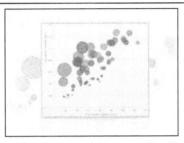 | **Bubbles**: These show a correlation between two to four different variables at once. The position in the X and Y axes gives the correlation between the first two metrics; the size and the color indicates the value of the third and fourth metrics. |
| 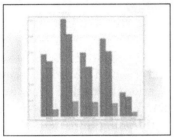 | **Bars**: These find relationships within your data very easily. They display your data as a vertical or horizontal graph, and view bars represent metric values of your data for each element of an attribute. |
| 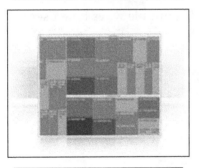 | **Heat/Tree Map**: These understand and compare a large number of values at once. They are a combination of colored rectangles organized in groups. The size represents the weighted value of a metric, and the color represents a relative change in another metric. |
| 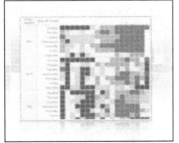 | **Matrix**: This graphically compares data across several variables. You can represent your data in several graphical forms, such as grid, scatter, bubble, line, or bar. Its main benefit is that you can add dimensionality to your graphs. |

| Visualizations | Description |
|---|---|
| 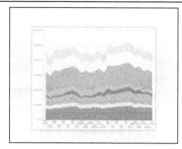 | **Area**: This connects points and shows changes in your data over time. This visualization allows you to see a contribution of an individual element to the total. This enables you to quickly analyze how the individual parts make up the whole. |
| 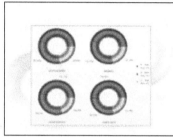 | **Pie**: This shows proportions of the parts of a whole. Focus on the size of each part as compared to the total. You should keep the number of the components to six or fewer, so as not to confuse the user with too many sections. |
| 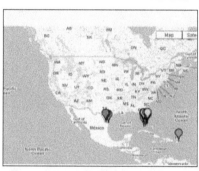 | **Maps**: These plot your analysis on a map using bubbles and pins to indicate location and performance by adjusting color and size. Drill down and select the bubbles or pins to investigate further (MicroStrategy provides ESRI and Google Map plugins to integrate map with your MicroStrategy data). |
| 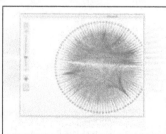 | **Network**: This organizes your data by nodes or a set of elements that are connected to each other by edges. In this visualization, the nodes represent the elements of a set, and the edges can be interpreted as a relationship between them. |

When activated, each visualization has its own configuration and properties. In the next section, we will learn how to configure and interact with visualization objects.

In our model, data metrics are loaded. We have already created five visualization objects without configuration under the *MicroStrategy interface – panels, layouts, and visualizations* section, now let's configure one of them:

1.  Click on the first visualization object located at the top-right corner of the screen.

2.  Select the **Heat Map** object.

3.  A new panel appears on the left side of the screen, next to dataset objects, for object configuration (in this case the heat map properties). Configure the properties according the following screenshot:

4.  A heat map is now visible. This map shows all the stores. The size of the box shows the profit by employee, the color denotes sales by employee (it is a threshold), and the tool tip shows the total sales at a particular data point on the heat map, which is a store in this case. Now the dashboard is similar to the following screenshot:

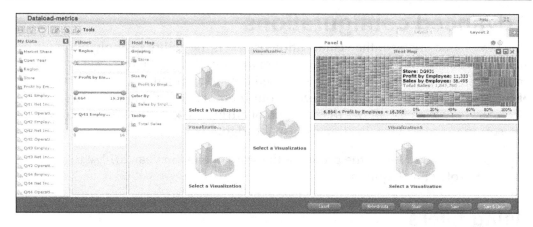

To complete this dashboard, repeat the preceding steps about selecting the visualization for each blank area on the dashboard (this is now applied to the object located in the top-left corner of the screen), but select the grid to add the visualization object. The grid panel is now activated.

Please be sure to select **Region** in the **Rows** field, no metrics in the **Columns** field, and **Total Sales** in the **Metrics** field. The grid now looks like the following screenshot:

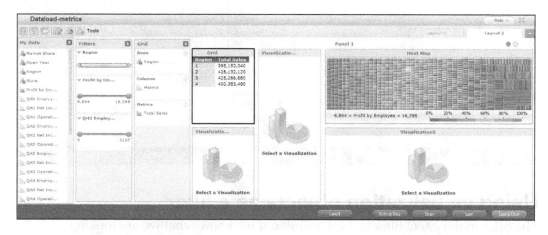

# Advanced configurations

One key characteristic of the visualization objects is the ability to apply filters and link visualization objects between them in order to find the root causes of the analysis. In the previous chapter, we learned how to apply filters, and the results of the filter will apply to all the objects of our report (grids or graphs). Also, latest MicroStrategy Web 9.4 allows the addition of attributes or metrics from multiple datasets in the same visualization in Visual Insight.

Suppose we need to know the profits of the store opened in year 2010, we have two options to solve this requirement.

## Using filters

In the filter panel, add a new filter, select **Open Year**, assure that the display type is a drop-down list, and select the year **2010**. Now, the visualization objects shows the total sales in the grid objects and stores opened in the selected year in the heat map, as shown in the following screenshot:

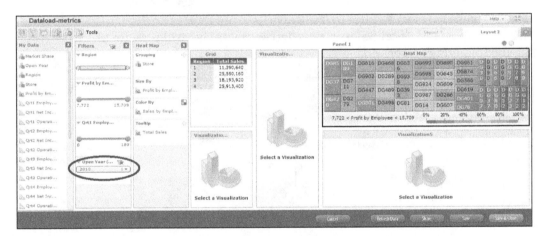

## Using visualization objects as a filter

The filter in the panel filter is a valid option if we know what we are looking for; but in some cases, we need to drill the information to discover the root cause of our analysis.

MicroStrategy functionality includes the ability to drill down the grid object and at the same time change other visualization objects in parallel, in order to show the results of the grid. In order to activate this functionality, follow the ensuing procedure:

1. Reset the open year filter (which was previously created) by clicking on the clear filter button [ 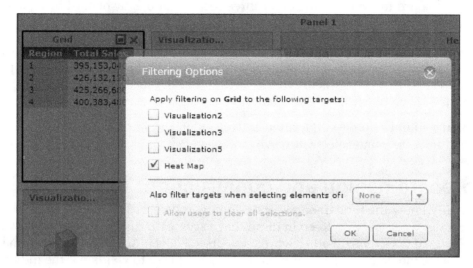 ] in the filter header, located in the filter panel.

2. Click on the drop-down button [ ] of the grid object and select **Use as a filter**. Select the heat map as the target, as shown in the following figure:

3. In the grid object of the region header, select drill to **Open Year**, scroll to the year **2010**, and drill to stores.

4. The grid will show only the total sales of the stores opened in year 2010, and the heat map only shows data from those stores. This is shown in the following screenshot:

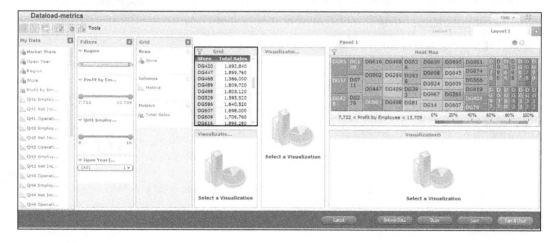

At the end of the day, the result is the same: the stores opened in year 2010, as a rule of thumb, reports displays only the data which customer wants to analyze and not the root causes.

You may follow the steps outlined previously while designing the first two sections of the dashboard to complete the other three sections of the dashboard by choosing the appropriate visualizations.

The best option is to use filters in the filter panel. On the other hand, if our report is focalized to analyze results, find the root causes and drill for details. The best option is to link visualization objects in order to drill information across all our report. Sometimes, when we drill for information, we need to start over. In order to do so,

click on the filter button [ ▽ ] located at the top-left corner of the header of each visualization object and select **Clear All**.

# Scorecard design and construction

We already know the fundamentals of designing a scorecard from the business and MicroStrategy perspective. Keep in mind that a scorecard displays progress towards a specific goal over time, in other words, the objective is to show the status of the key performance indicators in a given period of time in order to validate if the indicator is below or above the expected business results.

Prior to starting the design of the scorecard from the MicroStrategy standpoint, we need to identify the business issue that the scorecard will solve. In our case, to leverage the model that we have already loaded in MicroStrategy, the business issue is to track the profit of our stores by region, aligned to the thresholds defined by the commercial VP: red for profit below 450,000, yellow for a profit more than 450,000 but less than 550,000, and green for profits above 550,000.

In the first section of this chapter, we laid out several recommendations in order to design a successful scorecard. Now, the next step is to map the business issue to the best practices and align it as shown in the following table:

| Type | Strategies |
|---|---|
| Application emphasis | Measures performance |
| Users | Executives and management levels |
| Information | Summary with the option to drill down |
| Update frequency | Quarterly |
| Level of information | High level of profit, sales, employees, and profit by employee per store |
| Visual indicators | Heat map, bar charts, grids |
| Interactivity | Drills to store details, visual indicators are linked, offers filter to change regions and years when the store was open |
| Support for detecting trends and hunches | Yes |
| Shows stoplight indicators | Yes, defined by the commercial VP for profit by stores |

We have already designed the scorecard from the functional perspective; the next step is to start to building the scorecard in the MicroStrategy platform. In order to do so, we need to create a new model with the same raw data provided. This is done by following the ensuing procedure:

1. Log in to the MicroStrategy platform (or close the current model).

2. Generate a new model using the Excel file provided (`salesrawdatav1.0`), and click on the create visualization button [⊞] located on the main screen of MicroStrategy.

3. Confirm that only store, region, and open year columns are attributes (columns that organize and group metrics).

4. Select the grid option (you can change it later).

5. Click on the **Save** button. The system will request the name of the report, type `Scorecards - Dashboards`.

Now the data and layout of our model is ready. The next step is to build our reports.

The first step is to build the pieces of our scorecard like a lego, leverage the functional definitions already created, and then we will use those pieces to assemble the scorecard.

# Foundation

We need to define the foundation of our scorecard (the container of all the objects) based on the functional definition. We need only one layout and two panels; one for the status and resume of the information, and the second one for the drill operations. This is highly recommended in order to improve the usability of the scorecard.

> The drill function generates detailed information (columns and rows to scroll). In the scorecard section, the drilled results could be hard to read and understand. Hence, it is better to use additional panels to display detailed drill data.

Prepare the foundation by renaming the **Layout 1**, and change the title to Scorecard. Generate two panels, and rename them as Status and Details (new panel is added via the **Tools** menu located at the top left of the MicroStrategy interface). Then, remove the grid visualization object created by default. Now, the foundation is similar to the following screenshot:

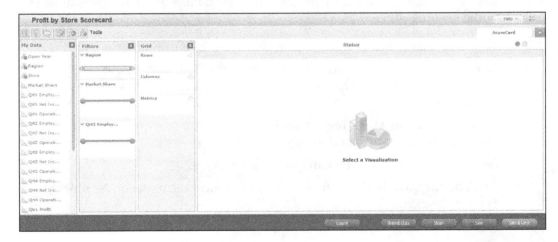

# Metrics

We need to generate metrics such as total sales, total employees, and total sales per store for our scorecard. Please refer to *Chapter 3, Reporting – from Excel to Intelligent Data*, if you have any doubts about metric generation.

The metrics consolidate the profits, sales, and employees by store. This information is more relevant for our scorecard and is the tip of the iceberg to start drilling and detecting root causes for specific store behaviors.

# Visualization objects

We need visualization objects for our scorecard. For Panel 1, it is an executive resume for which we will use the following objects:

| Panel 1 – Status | |
| --- | --- |
| Grid | Shows consolidate metrics by region |
| Heat map | Big picture of stores by region and total profit is the key metric to be analyzed |
| Matrix chart | Performance by region, profit, employees, and sales |
| **Panel 2 – Details** | |
| Grid | Detailed information per store |
| Bar graph | Detailed head count of the store |

# Building a scorecard

Now is time to start building the scorecard, let's start with the first panel named **Status**:

1. Add three visualizations objects and arrange them as shown in the following screenshot:

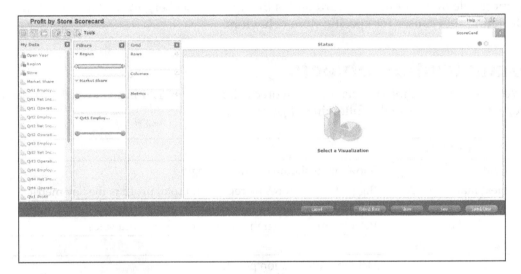

2. The first visualization object, which is located at the top left, is the most important one. Select a grid visualization and a grid panel will appear; ensure that only the total sales, total profit, and total employees appear in the metrics sections. Rename the grid object title to `Totals by Region`.

3. In the second visualization object, which is located at the top right, select a heat map object. A heat map panel will appear on your left. Select **Store** in the **Grouping** field and only **Total Sales** in the **Size** field.

4. Select **Total Profit** in the **Color By** field. In the **Total Profit** field, select the threshold according to the business rule defined by the commercial VP: red for profit below to 450,000, yellow for a profit more than 450,000 but less than 550,000, and green for profits above 550,000. The threshold looks like the following screenshot:

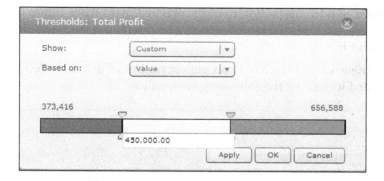

5. Rename the visualization object to Store Profit Heat Map. The scorecard now looks similar to the following screenshot:

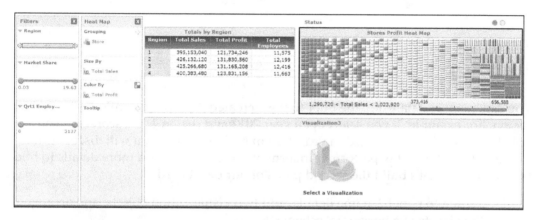

6. For the third visualization object, which is located at the bottom, select the matrix object. A graph matrix panel is activated. Configure it as follows:

| Attribute | Value |
| --- | --- |
| Rows | Metrics |
| Columns | Region |
| X axis | Total employees |
| Y axis | Total profit |
| Color by | Region |
| Size by | Total sales |
| Tooltip | Row count (total stores by region) |

7. Hide the legends of the graph and rename the visualization object to Sales, Employees and Profit by Region.

8.  We need to add interaction to our scorecard. Add three filters in the filter panel: one each for the region, store, and open year, and add the display type for each filter as a search box.

9.  Close the grid and the **My Data** panels for better usability. Now the scorecard looks like the following screenshot:

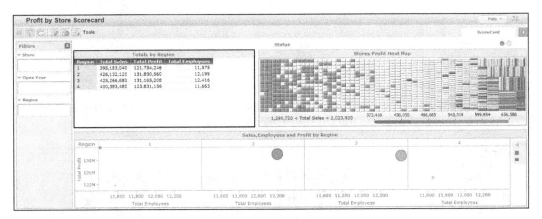

You can analyze data using the filter that we created: by store, specific open year, or region. For example, if you select open year 2012 and region 1, the scorecard shows only the total of stores opened in year 2012 in region 1. And, you will discover that the store DG689 is a low performer in term of revenue. We need more details to find the root cause. Let's build the second panel of our dashboard:

1.  Switch to panel 2, name details, add two visualization objects, and arrange the objects in a horizontal schema.

2.  For the first visualization object, which is located at the top, select grid. The grid panel appears. Configure it as follows:

| Attribute | Value |
| --- | --- |
| Rows | Stores |
| Columns | Metrics |
| Metrics | Sales, net income, operative income, profit for all the quarters |

3. Arrange the grid columns by quarter, sales, net income, operative income, and profit.

4. Format the data of each metric to currency, and rename the visualization object to `Store Details`. The grids now look like the following screenshot:

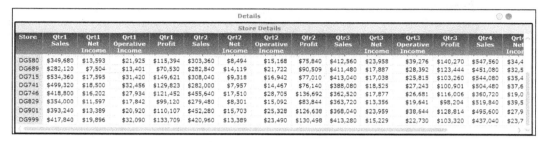

| Store | Qtr1 Sales | Qrt1 Net Income | Qrt1 Operative Income | Qtr1 Profit | Qtr2 Sales | Qtr2 Net Income | Qrt2 Operative Income | Qtr2 Profit | Qtr3 Sales | Qrt3 Net Income | Qrt3 Operative Income | Qtr3 Profit | Qtr4 Sales | Qrt4 Net Incor |
|---|---|---|---|---|---|---|---|---|---|---|---|---|---|---|
| DG580 | $349,680 | $13,593 | $21,925 | $115,394 | $303,360 | $8,494 | $15,168 | $75,840 | $412,560 | $23,958 | $39,276 | $140,270 | $547,560 | $34,4 |
| DG689 | $282,120 | $7,504 | $13,401 | $70,530 | $282,840 | $14,119 | $21,722 | $90,509 | $411,480 | $17,887 | $28,392 | $123,444 | $451,080 | $32,5 |
| DG715 | $534,360 | $17,595 | $31,420 | $149,621 | $308,040 | $9,318 | $16,942 | $77,010 | $413,040 | $17,038 | $25,815 | $103,260 | $544,080 | $35,4 |
| DG741 | $499,320 | $18,500 | $32,456 | $129,823 | $282,000 | $7,957 | $14,467 | $76,140 | $388,080 | $18,525 | $27,243 | $100,901 | $504,480 | $37,6 |
| DG746 | $418,800 | $16,202 | $27,934 | $121,452 | $455,640 | $17,510 | $28,705 | $136,692 | $362,520 | $17,877 | $26,681 | $116,006 | $360,720 | $19,0 |
| DG829 | $354,000 | $11,597 | $17,842 | $99,120 | $279,480 | $8,301 | $15,092 | $83,844 | $363,720 | $13,356 | $19,641 | $98,204 | $519,840 | $39,5 |
| DG901 | $393,240 | $13,389 | $20,920 | $110,107 | $452,280 | $15,703 | $25,328 | $126,638 | $368,040 | $23,959 | $38,644 | $128,814 | $495,600 | $27,9 |
| DG999 | $417,840 | $19,896 | $32,090 | $133,709 | $420,960 | $13,389 | $23,490 | $130,498 | $413,280 | $15,229 | $22,730 | $103,320 | $437,040 | $23,7 |

5. Configure the visualization object 2 by selecting vertical bar graph. The graph panel appears, and you can configure it as follows:

| Attribute | Value |
|---|---|
| Categories | Stores |
| Series | Metrics |
| Metrics | Qrt1 to Qrt4 employees |

6. Rename the visualization object to `Total Employees`.

You can arrange the metrics in chronological order using the drag-and-drop functionality. The same technique applies to other objects.

The scorecard is created. Close all of the panels except the filter panel, which always remains open and active when you switch the panels.

Type DG689 in the store filter and review the **Details** panel. You will see all the information related to that store in order to detect the root causes. This is shown in the following screenshot:

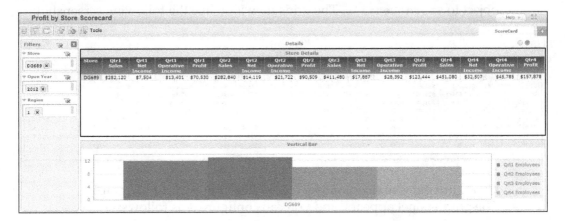

 Sometimes, when we drill for information, we need to start over. In order to do so, click on the filter button [ ⊽ ] located at the top left of the header of each visualization object and select **Clear all**.

Congratulations! Your first scorecard is created. As you can imagine, the combinations are endless for data analysis as scorecards leverages MicroStrategy platform.

You can use more filters, more visualization objects, more layouts and panels; but keep the design rules in mind. Overutilization of objects may cause a useless scorecard.

When building a dashboard from the platform perspective, the process to build a dashboard is quite similar to the process of building a scorecard. Define the foundation, metrics, and visualization objects. The main difference is the design scope from the functional standpoint.

Please keep in mind that a Business Intelligence dashboard, like the dashboard of a car that indicates the status of data at a specific point in time, is a snapshot of the current status versus business goals.

Now, we have the business need to generate a dashboard for our region managers to check the performance of their stores and region. The design principles of a dashboard will be as follows:

| Type | Operational/Tactical |
|------|----------------------|
| Application emphasis | Monitors operations, reviews progress per region |
| Users | Supervisors, office workers |
| Information | Summary/detailed |
| Update frequency | Intraday, daily/weekly |
| Level of information | Snapshots of key performance indicators, with options to drill |
| Visual indicators | Gauges, pies, matrices, and stacked bars |
| Interactivity | One screen of key indicators, with optional drill |
| Support for detect trends and hunches | Yes |
| Show stoplight indicators | Yes, linked to goals |

# Foundation

The foundation of our dashboard is based on one layout and one panel. Leverage our current model (scorecard or dashboard), add a new layout, and rename it to `Dashboard`.

# Metrics

We will leverage total sales, total employees, and total sales per store metrics already created in our model.

Now we need to create more metrics, such as sales by employee, profit by employee, and total operative costs (which is calculated by the total sales minus total profit). Please refer to *Chapter 3, Reporting – from Excel to Intelligent Data*, if you have any doubts about metric generation.

# Visualization objects

Our dashboard includes only one panel, with four visualization objects that are as follows:

| Panel 1: Dashboard | |
| --- | --- |
| Pie/donut | Market share of the region |
| Bar and area | Operative cost, total sales, and total profit |
| Stacked area | Sales and profit by employee |
| Matrix | Average sales and profit by store |

# Dashboard building process

As we already mentioned, the building process is quite similar to the process of scorecard creation. The final dashboard is shown in the following screenshot:

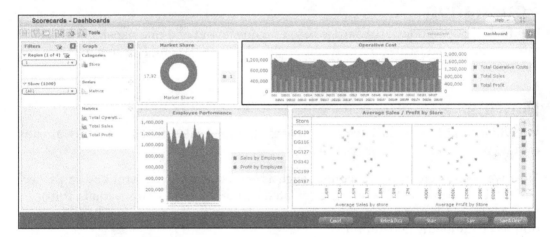

The properties of the visualization objects are located in *Appendix B, Visualization.*

# Summary

In this chapter, we learned how to transform our data into powerful scorecards and dashboards, from the business needs, functional design, and platform-driven configuration.

MicroStrategy offers a solid and easy-to-use platform. We highly recommend following the first steps prior to starting the configuration of the scorecard or dashboard: understand the business needs and follow the best practices. Dashboards and scorecards may turn out to be a very useful asset for organizations wanting to deploy a decision support system for any given business domain. Business domains such as financial and investment, insurance, healthcare, retail, and logistics may use the power of MicroStrategy Business Intelligence. Using this, executive users and business analysts can build dashboards and scorecards on their own. Business analysts can also share the reports with other business executives, or save it at the **Shared reports** location for use and collaboration by other users and analysts.

# 5
# Sharing Your BI Reports and Dashboards

The final objective of the information in the BI reports and dashboards is to detect the cause-effect business behavior and trends, and trigger actions to solve them. These actions supported by visual information, via scorecards and dashboards. This process requires an interaction with several people.

MicroStrategy includes the functionality to share our reports, scorecards, and dashboards, regardless of the location of the people.

## Reaching your audience

MicroStrategy offers the option to share our reports via different channels that leverage the latest social technologies that are already present in the marketplace, that is, MicroStrategy integrates with Twitter and Facebook.

The sharing is like avoiding any related costs and maintaining the design premise of the do-it-yourself approach without any help from specialized IT personnel.

# Main menu

The main menu of MicroStrategy shows a column named **Status**. When we click on that column, as shown in the following screenshot, the **Share** option appears:

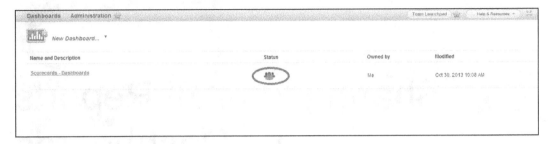

# The Share button

The other option is the **Share** button within our reports, that is, the view that we want to share. Select the **Share** button located at the bottom of the screen, as shown in the following screenshot:

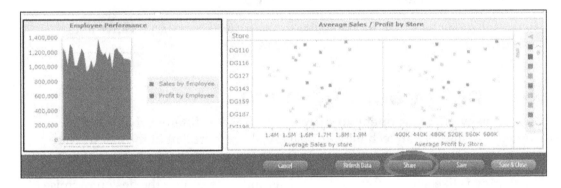

The share options are the same, regardless of the location where you activate the option; the various alternate menus are shown in the following screenshot:

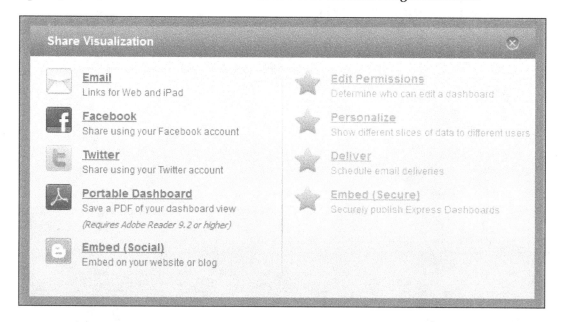

# E-mail sharing

While selecting the e-mail option from the **Scorecards-Dashboards** model, the system will ask you for the e-mail programs that you want to use in order to send an e-mail; in our case, we select Outlook.

MicroStrategy automatically prepares an e-mail with a link to share it. You can modify the text, and select the recipients of the e-mail, as shown in the following screenshot:

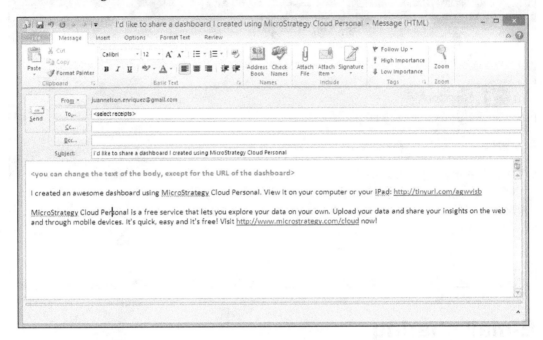

The recipients of the e-mail will click on the URL that is included in the e-mail, send it by this schema, and the user will be able to analyze the report in a read-only mode with only the **Filters** panel enabled.

The following screenshot shows how the user will review the report. Also, the user is not allowed to make any modifications.

This option does not require a MicroStrategy platform user account.

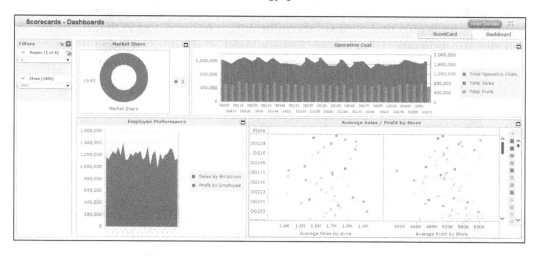

When a user clicks on the link, he is able to edit the filters and perform their analyses, as well as switch to any available layout, in our case, scorecards and dashboards. As a result, any visualization object can be maximized and minimized for better analysis, as shown in the following screenshot:

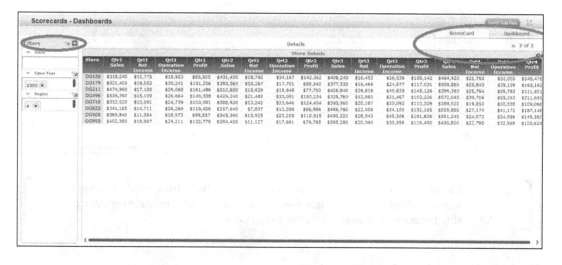

In this option, the report can be visualized in a fullscreen mode by clicking on the fullscreen button [  ] located at the top-right corner of the screen. In this sharing mode, the user is able to download the information in Excel and PDF formats for each visualization object. For instance, if you need all the data included in the grid for of the stores in region 1 opened in the year 2000. Perform the following steps:

1. In the browser, open the URL that is generated when you select the e-mail share option.

2. Select the **ScoreCard** tab.

3. In the **Open Year** filter, type 2012 and in the **Region** filter, type 1.

4. Now, maximize the grid.

5. Two icons will appear in the top-left corner of the screen: one for exporting the data to Excel and the other for exporting it to PDF for each visualization object, as shown in the following screenshot:

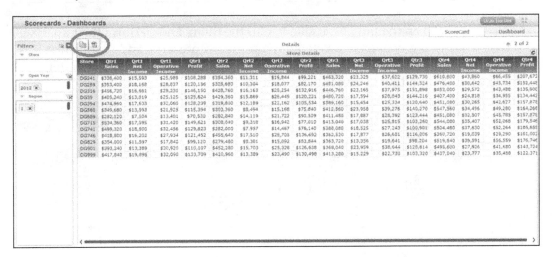

> Please keep in mind that these two export options only apply to a specific visualization object; it is not possible to export the complete report from this functionality that is offered to the consumer.

# Portable dashboard

The portable dashboard option will generate an interactive PDF (with flash technology embedded in it) that you can send via e-mail or analyze in an offline mode.

This option is very useful to share our reports with key directors when they are on the road or outside of their office, where there is no network connectivity. The following screenshot shows our model in the PDF reader:

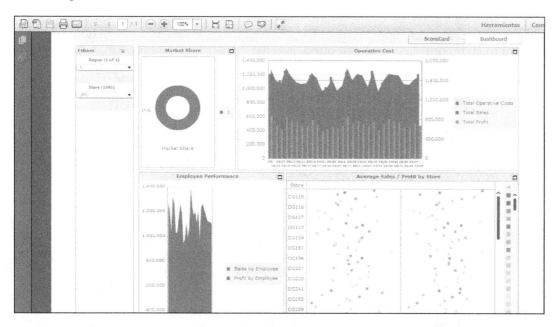

The objects in the PDF report are "clickable". You can select any layout, in our case, scorecards or dashboards, and maximize any visualization object that is available for better analysis.

The procedure to generate a portable dashboard is quite simple. Click on the share button [  ] from the main menu or the **Share** button if you are in the editing mode of the model, and select a portable dashboard option; the system will request you to save or open the created PDF.

This PDF can be shared via e-mail or in your intranet system. The required data is embedded in the file; therefore, a connection to the network is not required.

# Sharing on the social media – Facebook and Twitter

MicroStrategy allows you to share your insights with friends, colleagues, and followers on Facebook and Twitter. The MicroStrategy share options include the option to share our dashboard in our Facebook or Twitter timeline.

When we select this option, MicroStrategy accesses our Facebook and Twitter accounts (requires our authorization) in order to post the message.

## The Facebook option

The following screenshot shows the dialog screen that appears while selecting the Facebook option for sharing your insights:

# The Twitter option

The registration/sign-in dialog screen will appear while selecting the Twitter option for sharing the analysis, as shown in the following screenshot:

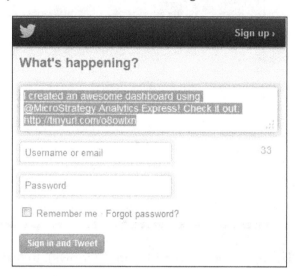

This option is very useful if the people who will review the report belong to our social networks, as well as when our report includes public information, and when we need to share the key figures with our audiences as part of a communication strategy.

# More Share Visualization options via the Web or user blogs

If we need to insert our dashboard in a specific web page or blog, MicroStrategy offers the option via HTML code.

When we select this option, MicroStrategy shows the code that is to be copied, as shown in the following screenshot:

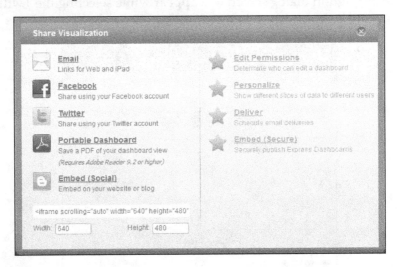

Select the code via the copy function (*Ctrl + C* in Windows). In our case, the code is as follows:

```
<iframe scrolling="auto" width="640" height="480" src="https://
my.microstrategy.com/MicroStrategy/servlet/mstrWeb?pg=shareAgent&doc
umentID=2CF2D93C11E260B8000000802FD7A04C&RRUid=1601634&starget=1"></
iframe>
```

In order to view the results, we generate a blog in the blogger platform (www. blogger.com,) and post the code generated by MicroStrategy. The results can be seen in the following screenshot:

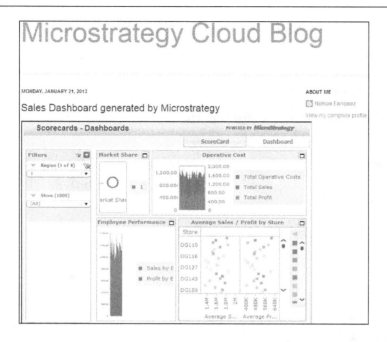

As you can see, the copied code refers to the MicroStrategy report that we share.

The objective of this option is to complement the report with more contextual and support information such as detailed sales reports and product portfolio.

One key advantage of the embedded (social) sharing option is the collaboration and feedback functionality that internal blogs or an organization's web application can provide. Users may give feedback or ask questions/clarifications using the organization's internal blog functionality. For instance, the commercial VP asks a question related to the performance in the region, and the regional manager responds with details using the organization's internal web application.

All the information logged in the conversation between the commercial VP and the regional manager is available for other users of the web application to review and further, comment if required under the same report. Here, it is important to discuss that an organization does not necessarily need to set up any such web application for collaboration, as the MicroStrategy web application also provides reports or document-reviewing capability in the application itself. This collaboration feature is available with the use of the **NOTES** feature in the document.

# Unshare the report

When we share a report via the different options that we already described, the security is controlled via the URL that MicroStrategy generates. This means that the security with this alternative has only two statuses: access or no access.

If we need to stop the sharing of our reports, that is, no access, we need to perform the following procedure:

1. In the main menu of the MicroStrategy interface, click on the desired report.

2. Select the **Unshare** option in the **Scorecards – Dashboards** panel as shown in the following screenshot:

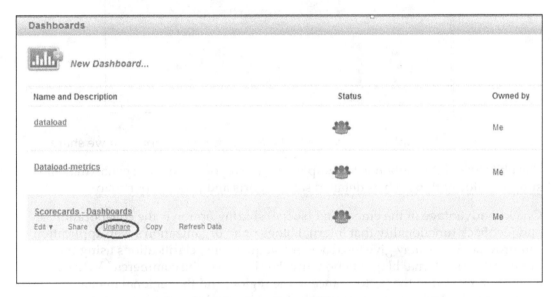

Now that the report is no longer shared, the URL sent via e-mail, Twitter, and Facebook, or embedded in our blogs is no longer valid, and the user will receive an error.

If you want to share the same report again, a new URL has to be generated; therefore, you need to send the e-mail, Twitter, and Facebook post again in order to activate it.

# Mobile devices

When we share a report, as we already mentioned, a web page URL is generated in order to be shared via several channels. The URL generated by MicroStrategy leverages the latest web technologies and is fully compatible with mobile devices having iOS and Android operating systems.

The following figure shows the URL on an Apple iPhone and an Android mobile device respectively:

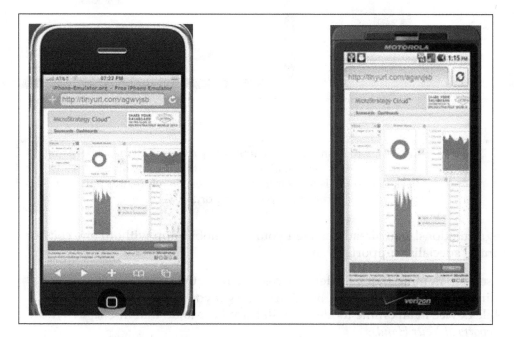

The following figure shows the URL on an Apple iPad and an Android tablet respectively:

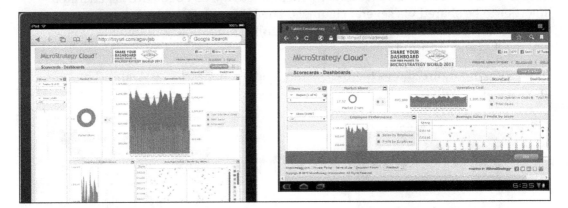

 The best way to share the MicroStrategy report with our users using mobile devices is via e-mail. We assume that the customers already check their e-mails in their mobile devices; therefore, it is only a one-touch process in order to open the report.

Congratulations! You already created your first mobile app without any code or technical specialist supporting you.

Please keep in mind that in order to access the report via a mobile device, you require an Internet connection. But don't worry if you need to work on your mobile device in an offline mode; we will learn how to do that in *Chapter 7, BI Reports at Your Hands*.

# Summary

In this chapter, we learned how to share our scorecards and dashboards via several channels such as e-mails, social networks (Twitter and Facebook), and blogs or corporate intranet sites.

The objective is to share our reports in order to trigger collaboration and gather knowledge for problem solving or root cause analysis for our business issues via a BI reports.

We discovered that it is very easy to share our reports in a mobile device or tablet; regardless of the technology vendor, the functionality is exactly the same. In fact, we can design our scorecards or dashboards for specific screen sizes for our mobile device, for example, arrange visualization objects in one single column instead of two or more to avoid scrolling in the screen. The only constraint of this option is the always-connected requirement of the device for the report access.

# 6
# MicroStrategy and the Cloud

In the previous chapters, we learned how to use the MicroStrategy platform for creating new models, reports, scorecards, and dashboards.

We also saw how to share via different channels without any complex setup of hardware, software, and specialized consultancy from an IT specialist; everything in a do-it-yourself approach.

All of this is possible due to the easy-to-use platform and, more importantly, the usage of the Cloud. But in order to be on the same page, we need to understand what the Cloud is.

In a Cloud approach the companies lease their digital assets, and their users are abstracted from the physical location of resources, such as data centers, applications, and databases they're using.

These resources are just in the Cloud somewhere; in our case, they are in the MicroStrategy Cloud. Firstly, in the Cloud approach, customers lease the infrastructure and software platform instead of buying them, shifting IT from a capital expense to an operating expense. This is essential for companies as they do not require huge investments in setting up the environment and maintaining them. Organizations can continue to focus on business as opposed to worrying about the environment.

Second, vendors are responsible for everything under the hood — all the maintenance, administration, capacity planning, troubleshooting, and backups.

And finally, the MicroStrategy Cloud platform is a high performance, flexible, reliable, and secure platform. The following table highlights some of the key characteristics of the MicroStrategy platform, which enables companies to choose to it.

| High performance | Flexible | Reliable and secure |
|---|---|---|
| <ul><li>64-bit architecture</li><li>In-memory computing</li><li>Best in class databases</li><li>Massive high-speed networks</li><li>State-of-the-art platforms</li></ul> | <ul><li>End-to-end managed services from development to user training</li><li>Connect directly to on-site data with MicroStrategy **Direct Connect** or host data in MicroStrategy Cloud</li><li>Analytical applications spanning all styles of Business Intelligence</li><li>Industry-leading BI for either web or mobile devices</li><li>Customize the interface by user and application</li></ul> | <ul><li>Systems:</li></ul>48 hours contract to development<br>High availability with 99.9 percent uptime guarantee<br>Highly redundant global infrastructure<br>Vulnerability and penetration testing<br>24 x 7 monitoring<ul><li>Architecture:</li></ul>Encryption<br>Customer isolation<br>Multifactor authentication<ul><li>Operations:</li></ul>Disaster recovery<br>24 x 7 operations team |

# The MicroStrategy Cloud portfolio

MicroStrategy offers three different versions of its platform in the Cloud, such as personal, express, and platform. They share the same concept of do-it-yourself schema and strong visualization capabilities, but with some extensions on the express and platform versions.

# Personal

This is the entry version of the offer provided by the MicroStrategy Cloud; it is easy to use within minutes and you are able to load data from your Excel files, generate a model, design reports, scorecards and/or dashboards, and share it all with your colleagues and customers. It is intended for only one person in the design and construction role.

This version is ideal for entrepreneurs, journalists, managers, professors, and anyone who needs to analyze data and communicate with their audience. Also, it gives 100 percent self-service and leverages the do-it-yourself approach.

You probably rely on spreadsheets to analyze and share data, as we already discussed. Almost everyone does, but the MicroStrategy Cloud personal edition is smarter. With just a few clicks, you can apply visual intelligence to your data; graph it, slice it, and filter it.

Moreover, this version includes the ability to share our models with mobile devices in a connected mode or offline mode.

The personal edition of MicroStrategy Cloud is free; no upfront or maintenance cost is required and the enablement is very easy, as explained in *Appendix A, MicroStrategy Express*.

# Personal edition

MicroStrategy Cloud personal edition characteristics are described in the following table:

| Characteristics | Description |
| --- | --- |
| Users | One (edition user) |
| Data volume | 1 GB |
| Data sources | MS Excel |
| Reporting | Data visualizations, interactive scorecards, and dashboards |
| Sharing | E-mail, social media, and blogs |
| Access mode | Web and iPad |
| Metadata | Automatic metadata creation. Access and analyze your data instantly. No data modeling or architecture necessary. |
| Setup | Instant |
| Tech support (in case you need it) | Discussion forum |
| Price | Free |

 This book and all the related examples and guidelines rely on the MicroStrategy Cloud personal edition. MicroStrategy 9.4 now calls the personal edition **Analytic Express**.

# Express

The express option includes all the functionalities and benefits of the personal edition. Additionally, it provides the design in advanced mobile interfaces without coding, access on premise and Cloud databases, delivery of personalized dashboards to more users, enterprise-grade security, and user management.

In fact, the security and on-premise database access functionalities are more beneficial for an enterprise use of the service.

## Express edition

The MicroStrategy Cloud express edition characteristics are described in the following table:

| Characteristics | Description |
|---|---|
| Users | Unlimited (the size of your wallet) |
| Data volume | 1 GB per user |
| Data sources | • MS Excel<br>• Files stored in Web locations<br>• Cloud-based or on-premises relational databases |
| Reporting | • Data visualizations<br>• Interactive dashboards<br>• Pixel-perfect reports and dashboards<br>• Unique dashboard widgets<br>• Advanced and predictive analytics |
| Sharing | • E-mail<br>• Social media and blogs<br>• Secure sharing with other users<br>• Scheduled report distribution |
| Access mode | • Web<br>• iPad |
| Metadata | • Automatic metadata creation<br>• Instant access and analysis of data<br>• No data modeling or architecture necessary |
| Setup | Instant |
| Tech support (in case you need it) | Live phone, e-mail, and web support |
| Price | Per user monthly subscription and capacity-based annual contract |

# Platform

The platform version of MicroStrategy Cloud is focalized for large deployments across companies with Business Intelligence solutions and leverages the Cloud benefits. It also supports several models from different processes across the company sharing a common set of data.

This version also includes a high-end database platform for extracting, transforming, and loading data from several sources.

The components of the solution are quite different in the personal and express versions. In fact you require more knowledge of Business Intelligence reports, design, and support from an IT specialist for the platform setup; the approach of do-it-yourself is no longer valid in this alternative.

We already know the fundamentals and benefits of the personal edition of MicroStrategy Cloud and the key advantages of the express edition, in particular the security, on-premise database, and the automatic distribution. In the next section of this chapter, we will learn the advantages of the express edition.

# Express edition in action

The first step is to enable the express evaluation version in our account that we have already created. We need to create a new team from the main menu of MicroStrategy; select the **New Team...** option and add a name and description in the **Name and Description** field, which in our case is **Commercial** as shown in the following screenshot:

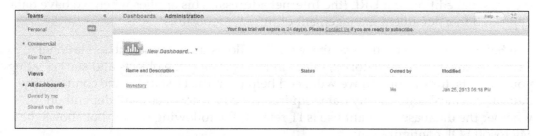

A team groups several reports of dashboards and scorecards. Imagine an area or business department within your company that is a team; each team will have its own security and controls.

# Data sources

Each team can include several reports, but the key difference in the creation of reports is the data source option. Let's create a new dashboard to review the options within the new team named **Commercial**, which is already created. Select a new dashboard and you will get the complete list of data source options, as shown in the following screenshot:

The data source for the Excel files includes the **Use File from Disk** option, similar to the personal edition or a URL (the Internet address). This applies when we have data deposited in a document management or intranet repository.

The **Salesforce** data and the **Database** option allows us to connect to a relational database, customers, inventories, sales, and so on. Select the objects and start creating your reports; for this option we will need help from the IT staff for the connectivity data. The typical connectivity data required for database access will depend on whether the database we want use is IT related; the following screenshot shows the Microsoft SQL configuration:

By default, when we select the database option, a **Sample Database** is loaded. Select this database and the following table by double-clicking on the **Credit_Card_ Purchases...** option and select all the related columns in the table by clicking on the add button [⊞] located at the right-hand side of the field, as shown in the following screenshot:

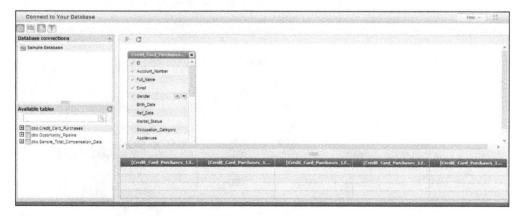

Then click on the **Continue** button and select the **Grid visualization** option; MicroStrategy will show the default view. Now, click on the **Save** button and name the report as Inventory Dashboard.

All the concepts that we learned about filters, visualization objects, layouts, and panels apply to this edition as well. In this case, the difference is the data source options.

# Security

Another value-added functionality in the express edition is security. With this version we can manage more users and also manage the level of access of each user to our reports. Let's start with users.

In the main menu of MicroStrategy, select the **Administration** option located at the top of the menu, as shown in the following screenshot:

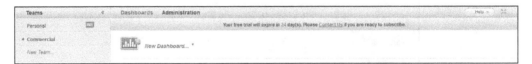

Now, the three main options, **Manage Users**, **Manager Groups**, and **Team Launchpad** (we will analyze later), will appear.

Select the **Manage Users** option and add a new user. The required data is the name, e-mail, and role such as **Administrator** (able to configure the platform and create new reports) or **Consumer** (is only able to view the information). Select the **Consumer** role; when the user is created in the system, it asks for user notification. Select **No** in this case.

For demo purposes, create a user with your alternate e-mail in order to play two roles.

The result will be similar to the following screenshot:

Now it is time to invite the new user to access the MicroStrategy platform; select the user and the message button [ ] and click on the **Send** option.

The new users will receive an e-mail with an invitation and a link to activate the account. When the user activates the account, the platform will request to generate a password and confirm the general data of the user.

The same procedure applies for the **Manage Groups** option. Select this option and then create a new group named **Consumers users**, and add the user that we created in the previous step.

 In order to add a user into a group, select the **Manage Groups** option in the user's list, click on the drop-down button [ ] located in the **Name** field of the user, and select the desired group.

## Applying security

When we have defined the users and groups, it is possible to apply security filters in our models for each user, in particular for two roles, **Can view** and **Can modify**.

This option is named **Set Permissions** and is located in the main dashboard view of the platform in the **Share** option of the model, as shown in the following screenshot:

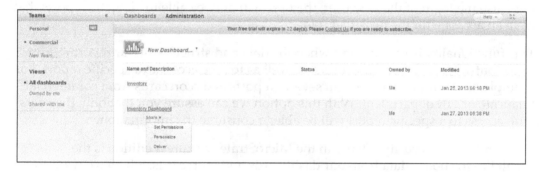

The system will show a screen in order to select a user (select the user that is already created) and assign the **Can view** permission in the **Inventory Dashboard** option, as shown in the following screenshot:

If we use the login of the new user, the user will only be able to view the reports that have an access with the designed role.

This functionality is very helpful when the data and structure of our report require to be controlled only for a few users, as well as to give access to a specific report. For example, the platform can contain several reports and scorecards from commercial, financial, or HR department. With this option we can assure you that only the users with access to a specific report will be able to consume the information.

Another level of security offered in the MicroStrategy express edition is the capability to choose data from our dashboards and scorecards, which we want to display to specific users. For example, we can display only the data for a specific region for one particular user even when the report contains data for all regions. This function is called personalize.

# Personalize

Using our inventory dashboard, let's configure the personalize option for the following case. The new user, which we have already created, needs to have access to the inventory information, but just for the **self-employee** customers.

In the administration view of MicroStrategy, the personalize attribute option is configured via the add button [ ] located in the **Manage Users** option.

Add a new personalization field named `customer type` with the data type text. For the new user, in the new personalization field, type `self-employees` as shown in the following screenshot:

Now in the main dashboard view, select the **Inventory Dashboard** option and map the personalization field customer type to the corresponding field in the inventory model (in the main dashboard option under the **Share** option).

Select the **Personalize** option, search for the **Occupation Category** attribute, select the personalization value **customer type**, and click on the **OK** button as shown in the following screenshot:

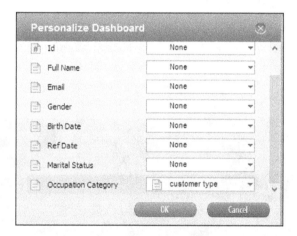

When the users with personalization fields access the dashboard, they only access the data where the attribute occupation category is equal to self-employees. This option is very useful when combined with the level of access to the dashboard. You can control which users will be able to change the report or only read it, and what content can be read within the report.

# Deliveries

Another key characteristic of the express edition is the ability to share our reports in an automatic way via a scheduled program. In the personal edition, we need to trigger the action manually in order to share the report.

This option only schedules the delivery via e-mail with a link, or embeds the PDF report within an automatic e-mail. The option to configure this functionality is in the main dashboard menu in the **Deliver** option under the **Share** menu, as shown in the previous section. The procedure to schedule a delivery is as follows:

1. Select the destination users from the register users list.
2. Define a subject for the e-mail.
3. Select the option to include the portable dashboard if it is required.
4. The body of the text is fully configurable. The text between {} are the system variables. Please don't modify their content; you can delete it if you want.

In the frequency selector, you can immediately select from a specific schedule as shown in the following screenshot.

- Daily
- Weekday
- Weekly
- Monthly

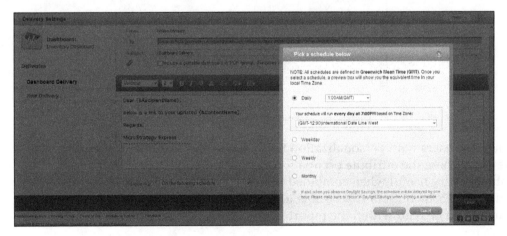

The system includes an option to deliver the report one time, immediately, besides the scheduled delivery time.

You can generate several deliveries of the same report with different users and schedules; for example, you only need the dashboard once a month for the Commercial VP, but you would need it every week for the store managers.

Also, it is possible to modify a predefined schedule via the deliveries menu or in the

main dashboard view by selecting the message button [<span></span>] located at the right-hand side of the desired model, as shown in the following screenshot:

## Refresh data

The refresh data functionality allows us to reload our data without compromising the design of the reports; we don't need to rebuild it when the data of the model changes.

In the personal edition, this option is triggered manually, but in the express edition this functionality can be automated if the data structure remains the same. If we have more columns or changes in data types, we need to reimport and analyze the data again; we don't need to rebuild our model even with this option.

This option is enabled in the main menu of the dashboards available in each report. The option is called **Refresh data** and it has two alternatives; **Refresh now** and **Set schedule refresh**.

The **Set Schedule refresh** option allows us to schedule a refreshing of the data at a specific point of time and is similar to the delivery options, as shown in the following screenshot:

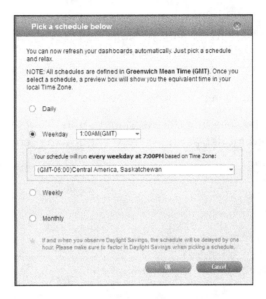

 If the data of the source changes in terms of structure, when the refresh data is executed, the design data load option is triggered in order to map the new fields.

# Summary

During this chapter we learned why the Cloud option is a real alternative for our reporting needs today, as well as we understood the complete portfolio that MicroStrategy offers to leverage the Cloud concept, in particular the express option, which has a cost.

The express option offers extra functionality over the personal option, such as access databases as the source of data, security for report access and attributes within the reports, automate report deliveries, and automate the refreshing of data when this data changes.

The personal edition functionality for designing and configuring the reports is exactly the same as that we used in the express edition. In MicroStrategy 9.4.1, the release Cloud platforms are now called analytics platforms.

# 7
# BI Reports at Your Hands

One advantage that MicroStrategy offers, is the ability to review our scorecards and dashboards from our mobile devices without an extra effort towards configuration.

The scope of mobile MicroStrategy includes mobile browser solutions and mobile BI client solutions. The mobile browser solutions enable the sharing of the reports, as we explained in *Chapter 5, Sharing Your BI Reports and Dashboards*, via web links (URLs), and then we can access the information using the web browser in the device. The reports work, but are not fully functional in terms of usability, and you require an Internet connection for accessing the report data.

An advanced option is to install a mobile BI client solution application in your device; this application is able to access our dashboards and scorecards in a more native way and is able to access the reports in the offline mode.

In order to enable this functionality, you need to download the application from the application market place for your mobile device (iTunes for Apple and Google play for Android), as shown in the following screenshot. These apps are available for free download, and any future updates to the application are pushed to the mobile device:

# The MicroStrategy Mobile BI client solution setup

When we download the app in the mobile device, the Apple iOS in our case, we only need to configure our username and password for the MicroStrategy platform, and you will able to access your reports immediately, as shown in the following screenshot:

All the security details and levels of access to our reports are maintained; the mobile apps of MicroStrategy leverage all the configuration settings for our users.

# MicroStrategy Mobile usability

When we install and set up the platform, the main menu of the mobile app will show your personal reports and shared reports, as shown in the following screenshot:

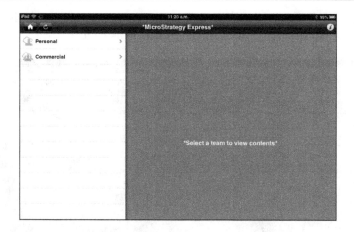

The personal reports that we created in the previous chapters are available to analyze and review in a native interface, which leverages the characteristics of the mobile device, in this case, the iPad.

We are able to navigate between panels and menus to drill data via selectors and filters previously created in our reports, as shown in the following screenshot:

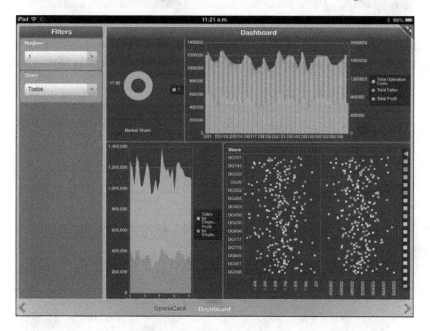

Selectors and graphs designed in the web reports are maintained in the mobile version without any changes. An example of this is as follows:

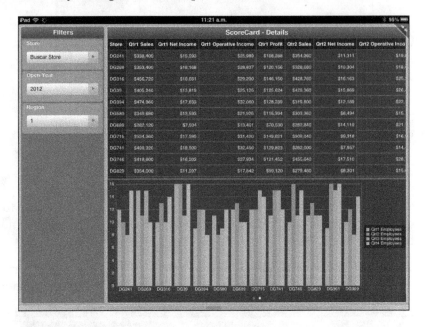

Grids, tables, and graphs are maintained in the same layout as defined earlier; the filters are totally functional and linked to the visualization objects. An example of this is as follows:

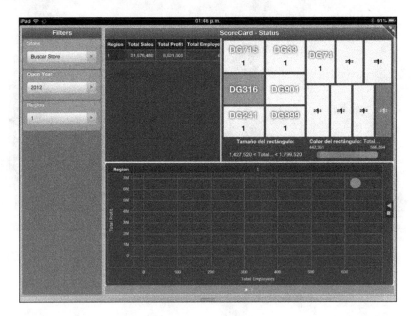

# The offline mode

One advantage of the MicroStrategy Mobile app is the ability to use the reports in the offline mode. This is enabled by default. You need to ensure that the report cache date is up-to-date, as shown here:

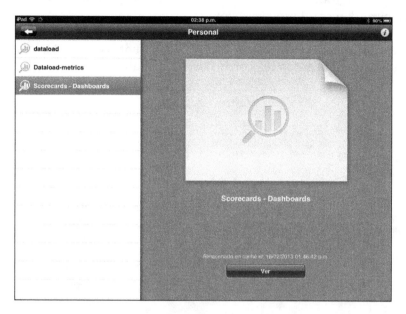

In order to assure that the cache is enabled and working for our reports, please be sure of the configurations in the advanced options of the mobile app configuration (the iOS option in this case):

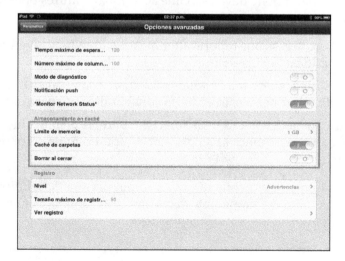

Other configuration settings can be kept default; these are controlled via the mobile device configuration on the mobile application server. Depending on the type of business, we may or may not expose these configuration settings for end users to change. User will not see the settings options if they are disabled in the mobile configuration on the mobile server. For more information on mobile configurations and general practices, you may refer to the MicroStrategy **Knowledge Base** and **Product Manual** sections and the **MicroStrategy Mobile Design and Administration Guide 9.4.1** PDF document. You need to choose the version and language to get to the correct list of available manuals:

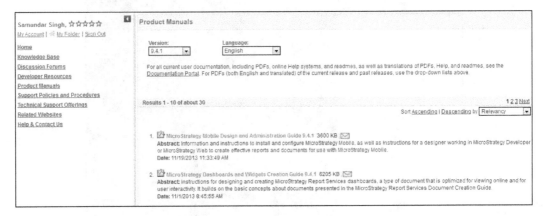

# Summary

One key characteristic of mobile BI is the fact that it is fully interactive with the BI content delivered to mobile devices. Some additional features of mobile BI are as follows:

- The ability to access our reports anytime and anywhere, even offline, in order to review critical business information and take actions

- The flexibility to share information via e-mail or messages from the mobile device with other people, in order to collaborate and share data

- The ability to navigate and drill down the BI reports in the mobile device with the same experience as a personal computer, in terms of usability and response time

The value proposition of the mobile BI offer is about pragmatic solutions: to deliver information needed to make immediate decisions with context, not just to push lots of data into the mobile device of the business user.

The MicroStrategy Mobile option is a flexible and easy-to-use alternative to enable a mobile BI strategy that enables mobile users to access our scorecards and dashboards via a mobile BI client solution, without any specific development or knowledge in the device platform. This functionality is a screen-driven process without the need for knowledge about mobile development platforms and device characteristics; it is a do-it-yourself schema that this offering of MicroStrategy is all about.

According to Aberdeen Group, Inc. (Aberdeen conducts primary research studies from a pool of over 500,000 panel participants. Aberdeen Group's research provides specific insight as per the industry sector, company size, and geography, as well as by job role, business process, and technology), a large number of companies are rapidly undertaking mobile BI due to tremendous market pressure on issues, such as the need for higher efficiency in business processes, improvement in employee productivity (for example, time spent looking for information), better and faster decision making, better customer service, and delivery of real-time, bidirectional data access to make decisions anytime and anywhere.

# A
# MicroStrategy Express

MicroStrategy Express is a hosted Cloud platform, and it is free for personal use. MicroStrategy also provides an enterprise version of the platform for user licensing and use, or a user may choose to set up on the premise Cloud.
More details on the latest offerings and news on these offerings can be found at `https://www.microstrategy.com/free/express`.

In this book, we will use free MicroStrategy Express to learn the MicroStrategy BI capabilities. In order to enable the MicroStrategy Cloud platform service, the only software you need are an Internet browser (Chrome, Firefox, or Internet Explorer) and an Internet connection. You can enable the MicroStrategy Cloud platform using the following the steps.

# Registering the Cloud service

In your web browser, access `http://www.microstrategy.com/express` and fill the **Sign in** and **Sign up** tabbed screens, as demonstrated in the following screenshot:

MicroStrategy Express Sign in and Sign up / registration page

Switch to the **Sign up** tab, type your full name, email address, select a password (your password must be at least eight characters long and should contain one of the following special characters: !, @, #, $, %, ^, &, +, =), and solve the mathematical problem (for human validation).

# Activating your account

You will receive an e-mail with a link for the account activation, similar to the following screenshot:

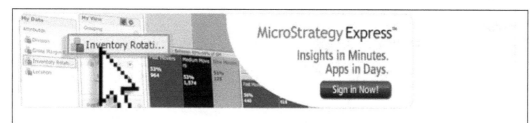

**Get Started Now!**

Hello Nelson,

Welcome to MicroStrategy Express! Click on the link below to verify your email address and activate your MicroStrategy Express account.

Click here to activate your MicroStrategy Express account.

Post-registration account activation page

Click on the link and the system will request more information about you, as shown in the following screenshot:

Welcome page

After finishing the capture, the system will redirect to the main page. Now you are set and ready to start working.

 The direct Internet address to access the system is `https://www.microstrategy.com/express/#sign-in`, and it is highly recommended to bookmark this address in your browser.

# B
# Visualization

## Visualization object's properties

Each visualization object has its own advanced properties. In order to activate them, click on the top-right corner of the object and select the **Edit Visualization...** option, as shown in the following screenshot:

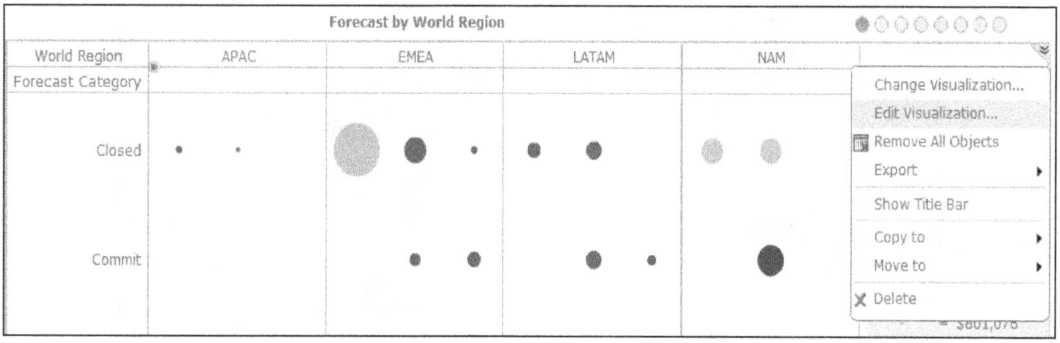

| World Region | APAC | EMEA | LATAM | NAM | |
|---|---|---|---|---|---|
| Forecast Category | | | | | Change Visualization... |
| | | | | | Edit Visualization... |
| Closed | • • | ● ● • ● ● | ● ● | | Remove All Objects |
| | | | | | Export ▶ |
| | | | | | Show Title Bar |
| Commit | | • ● ● • | | ● | Copy to ▶ |
| | | | | | Move to ▶ |
| | | | | | ✗ Delete |

Forecast by World Region

For objects such as charts, heat maps, lines, and stacks, a common set of properties are available:

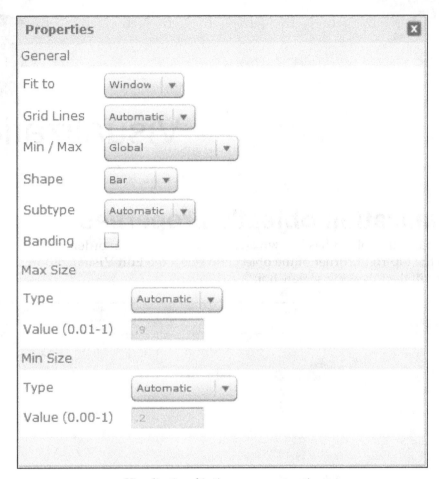

Visualization object's common properties

It is possible to adjust the fit of the object (`fit to`), hide or show the grid lines of the chart (`grid lines`), and choose the shape of the object from bar, line, circle, [insert square] and square (`shape`). It is also possible to manage the maximum size (`max size`) and minimum size (`min size`) of the objects.

# Visualization object's options

There are other available alternatives in the visualization objects that will allow us to change the object type, manipulate data, and copy or move the object to other panel. One useful option is the **Export** option (for exporting the object as an image), as shown in the following screenshot:

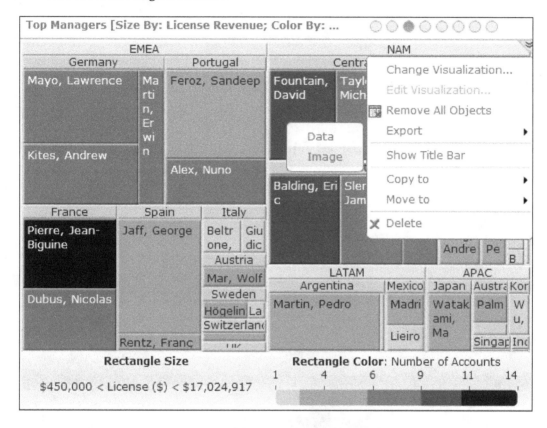

When you select this option, the visualization object is exported as a `.png` image for further use (send it by e-mail, use it in a PowerPoint presentation, and so on.)

Additionally, you can visit `http://www2.microstrategy.com/producthelp/` `9.3/WebUser/WebHelp/Lang_1033/About_visualizations.htm` to learn the details about different types of visualizations available for use and various visualization properties.

# Index

URL 115
**MicroStrategy Mobile app**
  setup 108
  usability 108-110
  usability, in offline mode 111, 112
**MicroStrategy Mobile BI client solution.** *See*
    **MicroStrategy Mobile app**
**mobile devices**
  used, for sharing reports 89, 90

# N

Network 59

# O

Online Analytical Processing (OLAP) 9

# P

panel 52
personalization, MicroStrategy Cloud
    express edition 102, 103
Pie 59
portable dashboard option 82, 83
predictive analytics 9

# R

Refresh data option 105,106
**reports**
  designing 21-28
  formatting 21-28
  sharing, mobile devices used 89, 90
  sharing, with MicroStrategy 77
  unsharing 88

# S

**scorecards**
  about 8
  versus, dashboards 48-50
**scorecards and dashboards**
  about 47
  benefits 48
  building 68-73
  building process 74
  configurations 62

designing 50-65
features 53-56
foundation 73
foundation, defining 66
layout 52
metrics 73
metrics, generating 67
panel 52
usability best practices 56
visualization 52
visualization objects 57, 67, 74
visualization, types 58-61
**security, MicroStrategy Cloud express**
    **edition**
  about 100, 101
  applying 101, 102
**Set Schedule refresh option 106**
**Share button 78**
**share visualization option 85-87**
**slices 37, 38**
**source data systems 11**

# T

**threshold 33-35**
**Twitter option 85**

# V

**value**
  searching, in data 39, 40
**visual insight 18**
**visualization**
  about 52
  types 58-61
**visualization objects**
  about 57, 67
  graph 18-20
  grid 18-20
  options 121
  properties 119, 120
  using, as filter 62-64

**Thank you for buying**
**Discovering Business Intelligence**
**Using MicroStrategy 9**

# About Packt Publishing

Packt, pronounced 'packed', published its first book "Mastering phpMyAdmin for Effective MySQL Management" in April 2004 and subsequently continued to specialize in publishing highly focused books on specific technologies and solutions.

Our books and publications share the experiences of your fellow IT professionals in adapting and customizing today's systems, applications, and frameworks. Our solution based books give you the knowledge and power to customize the software and technologies you're using to get the job done. Packt books are more specific and less general than the IT books you have seen in the past. Our unique business model allows us to bring you more focused information, giving you more of what you need to know, and less of what you don't.

Packt is a modern, yet unique publishing company, which focuses on producing quality, cutting-edge books for communities of developers, administrators, and newbies alike. For more information, please visit our website: www.packtpub.com.

# About Packt Enterprise

In 2010, Packt launched two new brands, Packt Enterprise and Packt Open Source, in order to continue its focus on specialization. This book is part of the Packt Enterprise brand, home to books published on enterprise software – software created by major vendors, including (but not limited to) IBM, Microsoft and Oracle, often for use in other corporations. Its titles will offer information relevant to a range of users of this software, including administrators, developers, architects, and end users.

# Writing for Packt

We welcome all inquiries from people who are interested in authoring. Book proposals should be sent to author@packtpub.com. If your book idea is still at an early stage and you would like to discuss it first before writing a formal book proposal, contact us; one of our commissioning editors will get in touch with you.

We're not just looking for published authors; if you have strong technical skills but no writing experience, our experienced editors can help you develop a writing career, or simply get some additional reward for your expertise.

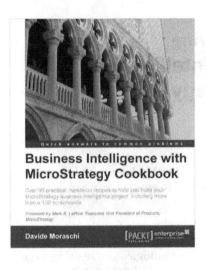

## Business Intelligence with MicroStrategy Cookbook

ISBN: 978-1-78217-975-7          Paperback: 356 pages

Over 90 practical, hands-on recipes to help you build your MicroStrategy business intelligence project, including more than a 100 screencasts

1.  Learn about every step of the BI project, starting from the installation of a sample database

2.  Design web reports and documents

3.  Configure, develop, and use the Mobile Dashboard

4.  Master data discovery with Visual Insight and MicroStrategy Cloud Express

## IBM Cognos Business Intelligence

ISBN: 978-1-84968-356-2          Paperback: 318 pages

Discover the practical approach to BI with IBM Cognos Business Intelligence

1.  Learn how to better administer your IBM Cognos 10 environment in order to improve productivity and efficiency.

2.  Empower your business with the latest Business Intelligence (BI) tools.

3.  Discover advanced tools and knowledge that can greatly improve daily tasks and analysis.

4.  Explore the new interfaces of IBM Cognos 10.

Please check **www.PacktPub.com** for information on our titles

**Oracle Business
Intelligence 11*g* R1
Cookbook**

Make complex analytical reports simple and deliver valuable
business data using OBIEE 11g with this comprehensive
and practical guide

Cuneyt Yilmaz     [PACKT] enterprise

# Oracle Business Intelligence 11g R1 Cookbook

ISBN: 978-1-84968-600-6       Paperback: 364 pages

Make complex analytical reports simple and deliver
valuable business data using OBIEE 11g with this
comprehensive and practical guide

1. Improve the productivity of business intelligence
   solution to satisfy business requirements with
   real-life scenarios

2. Practical guide on the implementation of OBIEE
   11g from A to Z including best practices

3. Full of useful instructions that can be easily
   adapted to build better business intelligence
   solutions

**Business Intelligence Cookbook:
A Project Lifecycle Approach
Using Oracle Technology**

Over 80 quick and advanced recipes that focus on real-world
techniques and solutions to manage, design, and build data
warehouse and business intelligence projects

John Heaton     [PACKT] enterprise

# Business Intelligence Cookbook: A Project Lifecycle Approach Using Oracle Technology

ISBN: 978-1-84968-548-1       Paperback: 368 pages

Over 80 quick and advanced recipes that focus on
real-world techniques and solutions to manage,
design, and build data warehouse and business
intelligence projects

1. Full of illustrations, diagrams, and tips with
   clear step-by-step instructions and real time
   examples to perform key steps and functions on
   your project

2. Practical ways to estimate the effort of a data
   warehouse solution based on a standard work
   breakdown structure.

3. Learn to effectively turn the project from
   development to a live solution

Please check **www.PacktPub.com** for information on our titles

www.ingramcontent.com/pod-product-compliance
Lightning Source LLC
Chambersburg PA
CBHW060149060326
40690CB00018B/4043